8 Days to Master

Fourier Series

without calculus

This page intentionally left blank

Humbert Cole

8 Days to Master

Fourier Series

without calculus

a step-by-step tutorial
for absolute beginners

Contents

Preface

The purpose of this book is to give the reader a solid introduction to Fourier series without a calculus prerequisite. It is intended for self-study.

This book simplifies Fourier series and introduces the concepts to the reader in a manner that eases understanding. It establishes the idea of Fourier series independent of calculus, thereby presenting a fresh and mathematically rich view of the subject.

Benefits of Reading this Book

- Get a good introduction to Fourier series.

- Learn a clean, fast and efficient method of solving Fourier series problems.

- Have a deep and conceptual understanding of Fourier series.

Target Audience   This book is for anyone who needs an introduction to Fourier series and has no prior knowledge of calculus.

Problems Solved by this Book

- The problem of calculus being a prerequisite to learn Fourier series.

- The problem of inefficient methods of solving Fourier series problems by hand. These methods involve writing a lot of symbols and carrying out integration several times to solve a single problem.

- Most textbooks do not consider the struggle of students and therefore fail to explain the core concepts of Fourier series properly.

Advice to the Reader

- Always make notes while reading. Take note of important formulas and theorems. Refer back to your notes to get better understanding.

- Do not rush when solving problems. Stay focused. Take your time. Speed comes naturally without you even thinking about it.

- Ensure that you work through the exercises. Check your answers in the *Full Worked Solutions to Exercises* section at the end of the book. The workings are also present there.

Humbert Cole

Introduction

What is Fourier Series?   Fourier series is a way of representing a periodic function as a sum of cosine and sine functions. The process involves breaking up a function into simpler bits that are easier to work with in real life applications.

History of Fourier Series

Fig. 1: Portrait of Joseph Fourier

Fourier series is named in honor of Jean-Baptiste Joseph Fourier (1768–1830). He was a French mathematician and physicist. Fourier established that an arbitrary continuous function could be represented by a series of sine and cosine functions. He introduced the series to solve the problem of heat distribution in a metal plate. Today, this series is applied to a wide array of mathematical problems. It has applications in electrical engineering, signal processing, vibration analysis, acoustics, etc.

Why is Fourier Series Important?

Fourier series plays a fundamental role in modern technology. We use it to process sounds, information and images. The music and video industry relies on Fourier series.

Below is an outline of the uses of Fourier series in various fields of science and engineering.

- **Signal Processing:** Fourier series is used to expand a periodic signal as a sum of various frequency components. It is essential in order to understand how signals behave when they pass through amplifiers, filters, etc.

- **Sound Processing:** A sound can be represented by a mathematical function with time as the free variable. It can be modeled as a sum of frequency components and decomposed into basis sounds to make its frequency content available. This decomposition is enabled by Fourier series.

- **Civil Engineering:** Fourier series is used to study vibrations in buildings. This is used to estimate the strength of structures.

- **Heat Conduction:** Fourier series can be used to solve the empirical relationship between the conduction rate in a material and the temperature gradient in the direction of energy flow.

- **Communication Theory:** In communication theory, the signal is usually a voltage. Fourier series is used to model how the signal behaves.

  Fourier series is also used in advanced noise cancellation for communication devices.

- **Solution of Partial Differential Equations:** Fourier series is used in the solution of partial differential equations which appear frequently in many mechanical engineering problems.

CHAPTER 1

Prerequisites

This chapter discusses the prerequisites for learning Fourier series with this book. It touches on sequences, trigonometric functions and infinite series among other topics.

1.1 Important Sequences

The table below shows the most common sequences encountered in the study of Fourier series.

Table 1.1

	Sequence	Description	Formula
1.	1, 2, 3, 4, ...	Natural Numbers	n
2.	1, 3, 5, 7, ...	Positive Odd Integers	$2n - 1$
3.	2, 4, 6, 8, ...	Positive Even Integers	$2n$
4.	1, 5, 9, 13, ...	Common difference is 4. Starting number is 1	$4n - 3$
5.	2, 6, 10, 14, ...	Common difference is 4. Starting number is 2	$4n - 2$
6.	3, 7, 11, 15, ...	Common difference is 4. Starting number is 3	$4n - 1$
7.	4, 8, 12, 16, ...	Common difference is 4. Starting number is 4	$4n$

1.2   Basic Properties of Trigonometric Functions

Let n be an integer, $n = 1, 2, 3, \ldots$, then

1.

$\sin(-nx) = -\sin nx$

$\cos(-nx) = \cos nx$

where x is a real number.

2.

$\sin n\pi = 0$

$$\cos n\pi = \begin{cases} -1 & n \text{ is odd} \\ 1 & n \text{ is even} \end{cases}$$

3.

$$\sin \frac{n\pi}{2} = \begin{cases} 0 & n \text{ is even} \\ 1 & n = 1, 5, 9, \ldots \\ -1 & n = 3, 7, 11, \ldots \end{cases}$$

$$\cos \frac{n\pi}{2} = \begin{cases} 0 & n \text{ is odd} \\ -1 & n = 2, 6, 10, \ldots \\ 1 & n = 4, 8, 12, \ldots \end{cases}$$

Do you wish to see how all these properties come about? Check Appendix B for a detailed discussion on how to derive these results from the sine and cosine functions.

Exercise 1

Evaluate the following.

1. $\sin 2n\pi$

2. $\sin 11n\pi$

3. $\cos 2n\pi$

4. $\cos 12n\pi$

5. $\sin n\pi + \cos 6n\pi$

Exercise 2

Evaluate the following.

1. $1 - \cos n\pi$

2. $\dfrac{1}{n^2} \sin n\pi$

3. $\dfrac{1}{n\pi}(\cos n\pi - 1)$

4. $\cos n\pi - \cos \dfrac{n\pi}{2}$

5. $\dfrac{1}{n}(\cos \dfrac{n\pi}{2} - \cos n\pi)$

6. $-\dfrac{2\pi}{n} \cos 2n\pi + \dfrac{1}{n^2} \sin 2n\pi$

7. $\dfrac{8}{n\pi} \sin \dfrac{n\pi}{2}$

8. $-\dfrac{4}{n\pi} \cos \dfrac{n\pi}{2}$

9. $-\dfrac{2}{n\pi} \sin \dfrac{n\pi}{2}$

10. $-\dfrac{4}{n^2\pi^2}(\cos n\pi - 1)$

11. $\dfrac{2}{n\pi} \sin \dfrac{n\pi}{2} + \dfrac{4}{n^2\pi^2}(\cos n\pi - 1)$

12. $\dfrac{2}{n^2\pi}(2\cos \dfrac{n\pi}{2} - \cos n\pi - 1)$

13.

$$\dfrac{1}{n}(1 - \cos n\pi) - \dfrac{\pi}{n} \cos n\pi$$
$$+ \dfrac{1}{n^2} \sin n\pi$$

1.3   The Sigma Notation

$$\sum$$

The sigma notation is very important to precisely write a Fourier series. It is a notation used to sum things up. $\sum n$ means "sum of n"

Now we need to know what values of n we are to sum up. The values are shown below and above the sigma.

$$\sum_{n=1}^{5} n$$

This means start from 1 and sum up till 5.

Thus

$$\sum_{n=1}^{5} n = 1 + 2 + 3 + 4 + 5$$

$$= 15$$

The summary of the rules of the sigma notation is shown below.

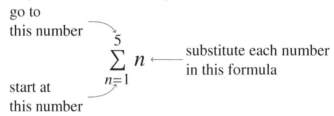

Example 1

Find

$$\sum_{n=1}^{4} (n + 1)$$

Solution

$$\sum_{n=1}^{4} (n + 1)$$

We are to sum $(n+1)$ and the value of n goes from $n = 1$ to $n = 4$. Therefore

$$\sum_{n=1}^{4} (n + 1) = (1 + 1) + (2 + 1) + (3 + 1) + (4 + 1)$$

$$= 2 + 3 + 4 + 5$$

$$= 14$$

Example 2

Find

$$\sum_{n=2}^{6} n^2$$

Solution

$$\sum_{n=2}^{6} n^2$$

We are to sum n^2 and the value of n goes from $n = 2$ to $n = 6$. Therefore

$$\sum_{n=2}^{6} n^2 = 2^2 + 3^2 + 4^2 + 5^2 + 6^2$$
$$= 4 + 9 + 16 + 25 + 36$$
$$= 90$$

1.4   Infinite Series

An infinite series is the sum of infinitely many terms. The terms could be numbers or functions. We are interested in them because Fourier series is an infinite series of sine and cosine functions.

1.4.1   Examples of Infinite series

1. Sum of all natural numbers   Remember our sigma notation? We can use it here. Sum of all natural numbers can be written as:

$$1 + 2 + 3 + 4 + \cdots = \sum_{n=1}^{\infty} n$$

The dots $\cdots$ mean that the sum continues forever. Notice that unlike previous examples, the sigma notation has infinity, ∞, as the 'number' above it. This is a way to identify infinite series. They always have ∞ above the sigma.

2. Sum of all positive powers of 2   Sum of all positive powers of 2 can be written as:

$$2 + 2^2 + 2^3 + 2^4 + \cdots = \sum_{n=1}^{\infty} 2^n$$

3. Sum of all positive powers of x   This is simply a generalization of the previous example. Sum of all positive powers of x can be written as:

$$x + x^2 + x^3 + x^4 + \cdots = \sum_{n=1}^{\infty} x^n$$

4. $\cos x + \cos 2x + \cos 3x + \cos 4x + \cdots$ This sum can be written as:

$$\cos x + \cos 2x + \cos 3x + \cos 4x + \cdots = \sum_{n=1}^{\infty} \cos nx$$

1.4.2   Partial Sums

Consider the infinite series

$$1 + 2 + 3 + 4 + \cdots$$

We know that

$$1 + 2 + 3 + 4 + \cdots = \sum_{n=1}^{\infty} n$$

We may decide to sum up the natural numbers up to a certain point. This is called a *partial sum.*

Let us call the last number k. Thus the partial sum is:

$$1 + 2 + 3 + 4 + \cdots + k = \sum_{n=1}^{k} n$$

(Here, the $\cdots$ means there are some terms between 4 and k)

Let us use the notation S_k to denote the partial sum up to the number k. Thus

$$S_k = \sum_{n=1}^{k} n$$

Therefore,

$$S_1 = 1$$
$$S_2 = 1 + 2 = 3$$
$$S_3 = 1 + 2 + 3 = 6$$
$$S_4 = 1 + 2 + 3 + 4 = 10$$
$$\vdots$$
$$S_{100} = 1 + 2 + 3 + \cdots + 100 = 5050$$

1.4.3 Convergence and Divergence

Sometimes, an infinite series sums up to a particular finite value. When this happens we say the infinite series *converges*. If it does not we say it *diverges*.

Let S_k be the partial sum of an infinite series up to the kth term. If S_k approaches a fixed number as k becomes larger, the series is said to converge.

If S_k grows larger without bound as k becomes larger, the series is said to diverge.

1.4.4 An Example of Convergence

Consider the infinite series

$$\frac{1}{1^2} + \frac{1}{2^2} + \frac{1}{3^2} + \frac{1}{4^2} + \cdots$$

The mathematician Euler proved that

$$\frac{1}{1^2} + \frac{1}{2^2} + \frac{1}{3^2} + \frac{1}{4^2} + \cdots = \frac{\pi^2}{6} \tag{1.1}$$

Therefore this is a convergent series which converges to the fixed value $\frac{\pi^2}{6}$. We can rewrite (1.1) as;

$$\sum_{n=1}^{\infty} \frac{1}{n^2} = \frac{\pi^2}{6}$$

Now let us take a partial sum of the infinite series up to the kth term. We can write this as

$$S_k = \sum_{n=1}^{k} \frac{1}{n^2}$$

In the Table 1.2, we shall observe how S_k gets closer and closer to $\dfrac{\pi^2}{6}$ as k becomes larger.

Table 1.2

$\dfrac{\pi^2}{6} \approx 1.6449$	

k	S_k
1	1.0000
10	1.5498
100	1.6350
1000	1.6439
2000	1.6444

We conclude that as k approaches ∞, S_k approaches $\dfrac{\pi^2}{6}$.

Exercise 3

1. If $a_n = \dfrac{8}{n\pi} \sin \dfrac{n\pi}{2}$, evaluate the sum $S = \displaystyle\sum_{n=1}^{\infty} a_n \cos nx$

2. If $b_n = -\dfrac{4}{n^2\pi^2}(\cos n\pi - 1)$, evaluate the sum $S = \displaystyle\sum_{n=1}^{\infty} a_n \sin nx$

3. If $a_n = \dfrac{1}{n\pi}(\cos n\pi - 1)$ and $b_n = -\dfrac{4}{n\pi}\cos\dfrac{n\pi}{2}$, evaluate the sum $S = \displaystyle\sum_{n=1}^{\infty} a_n \cos nx + \sum_{n=1}^{\infty} b_n \sin nx$

1.5   How to Interpret Infinite Series in this Book

There are seven sequences used in this book. See Table 1.1. All infinite series that appear in this book use one or more of these sequences. You can determine which sequence is present in an infinite series from the first two terms only.

Example 1

An infinite series is written as

$$\frac{1}{1^2} + \frac{1}{5^2} + \cdots$$

Judging from the first two terms, the sequence used in this series is

$$1, 5, 9, 13, \ldots$$

That is the $4n - 3$ sequence. Thus we can write the series with more terms as

$$\frac{1}{1^2} + \frac{1}{5^2} + \frac{1}{9^2} + \frac{1}{11^2} + \cdots$$

Example 2

An infinite series is written as

$$\frac{1}{2} \cos 2x + \frac{1}{4} \cos 4x + \cdots$$

Write more terms of the series.

Solution   The sequence used in this series is

$$2, 4, 6, 8, \ldots$$

That is the positive even integers. This sequence can be found in two locations:

- The denominator of the fractions

- The argument of the cosine functions

Thus we can write the series with more terms as

$$\frac{1}{2} \cos 2x + \frac{1}{4} \cos 4x + \frac{1}{6} \cos 6x + \frac{1}{8} \cos 8x + \cdots$$

Example 3

An infinite series is written as

$$\sin x + \frac{1}{2}\sin 2x + \cdots$$

Write more terms of the series.

Solution   There is an invisible $\frac{1}{1}$ at the start of the series:

$$\frac{1}{1}\sin x + \frac{1}{2}\sin 2x + \cdots$$

Now it is clear that the sequence used in this series is

$$1, 2, 3, 4, \ldots$$

That is the natural numbers. It can be found in two locations:

- The denominator of the fractions

- The argument of the sine functions

Thus we can write the series with more terms as

$$\sin x + \frac{1}{2}\sin 2x + \frac{1}{3}\sin 3x + \frac{1}{4}\sin 4x + \cdots$$

1.5.1 Alternating Series

This is a series whereby the sign of terms alternate between positive and negative. Examples are:

1. $1 - 2 + 3 - 4 + 5 - 6 + \cdots$

2. $1 - \frac{1}{3} + \frac{1}{5} - \frac{1}{7} + \cdots$

Example

An infinite series is written as

$$S = \frac{1}{2^2}\cos 2x - \frac{1}{6^2}\cos 6x + \cdots$$

Write more terms of the series.

Solution   The first term has a positive sign while the second term has a negative sign. Thus this is an alternating series. The sequence used is

$$2, 6, 10, 14, \ldots$$

That is the $4n - 2$ sequence. Therefore we can write more terms of the series as

$$S = \frac{1}{2^2}\cos 2x - \frac{1}{6^2}\cos 6x + \frac{1}{10^2}\cos 10x - \frac{1}{14^2}\cos 14x + \cdots$$

1.6   Area under a Curve

This section discusses certain rules for the area under the curve of a function $f(x)$.

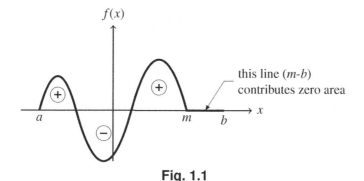

Fig. 1.1

The area under the curve of $f(x)$ between $x = a$ and $x = b$ is shown in Fig. 1.1. In order to find such area there are some rules we must apply.
The rules are;

1. Areas above the x-axis are positive.

2. Areas below the x-axis are negative.

3. Area between a horizontal line and the *x*-axis is zero *if the line is on the x-axis.*

Example

Find the area under the curve of $f(x)$ between $x = a$ and $x = b$ provided that $A_1 = 5$, $A_2 = 10$ and $A_3 = 15$ square units.

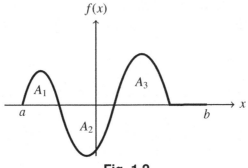

Fig. 1.2

Solution   The area A between $x = a$ and $x = b$ is;

$$A = A_1 - A_2 + A_3$$
$$= 5 - 10 + 15$$
$$= 10 \text{ square units}$$

CHAPTER 2

Periodic Piecewise Functions

This chapter introduces the concept of piecewise and periodic functions. This is the most common type of function modeled with Fourier series.

We begin the chapter by studying the equation of a straight line as this will be required to understand piecewise linear functions, which come later in the chapter.

2.1 Equation of a Straight Line Passing Through Two Given Points

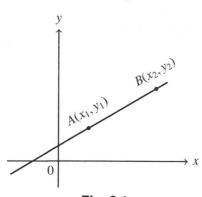

Fig. 2.1

Let two points A and B have the coordinates (x_1, y_1) and (x_2, y_2) respectively. The equation of the line passing through both points is given by:

$$y = \left(\frac{y_1 - y_2}{x_1 - x_2} \right)(x - x_1) + y_1$$

2.2 Piecewise Functions

A piecewise function $f(x)$ is one defined by multiple formulas, each formula applying to a certain section of the x-axis. In Fig. 2.2 we see a function made up of three pieces. Each piece has its own formula.

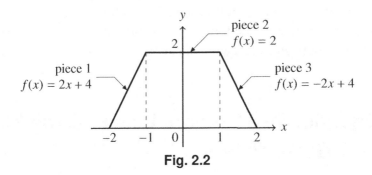

Fig. 2.2

In this case all the pieces are straight lines. This type of piecewise function is known as a *piecewise linear function*.

From Fig. 2.2 we can arrange the formulas for the pieces in an orderly manner. See Table 2.1.

Table 2.1

formula	section of x-axis
$2x + 4$	$-2 < x < -1$
2	$-1 < x < 1$
$-2x + 4$	$1 < x < 2$

Just like the table above, the standard notation for a piecewise function like $f(x)$ is:

$$f(x) = \begin{cases} 2x + 4 & -2 < x < -1 \\ 2 & -1 < x < 1 \\ -2x + 4 & 1 < x < 2 \end{cases}$$

Study the following examples carefully.

Example 1

Draw the graph of the following piecewise function.

$$f(x) = \begin{cases} -5 & -1 < x < 0 \\ 5 & 0 < x < 1 \end{cases}$$

Solution

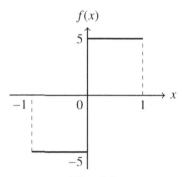

Fig. 2.3

Example 2

Draw the graph of the following piecewise function.

$$f(x) = \begin{cases} x + 1 & -\pi < x < 0 \\ x - 1 & 0 < x < \pi \end{cases}$$

Solution

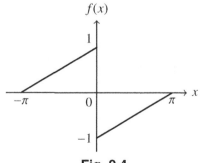

Fig. 2.4

Example 3

Determine the piecewise definition of the function shown below.

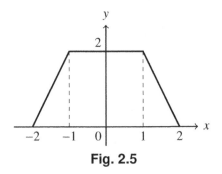

Fig. 2.5

Solution

The function is made up of three pieces which are all straight lines: line 1, line 2 and line 3. See the figure below.

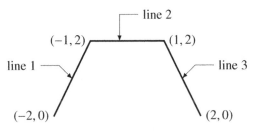

Fig. 2.6

Line 1   This line passes through $(-2, 0)$ and $(-1, 2)$.

$$y = \left(\frac{y_1 - y_2}{x_1 - x_2}\right)(x - x_1) + y_1$$

$$= \left[\frac{0 - 2}{-2 - (-1)}\right][x - (-2)] + 0$$

$$= \frac{-2}{-1}(x + 2)$$

$$= 2x + 4$$

Line 2   This line is an horizontal line.

$$y = 2$$

Line 3   This line passes through $(1, 2)$ and $(2, 0)$.

$$y = \left(\frac{y_1 - y_2}{x_1 - x_2}\right)(x - x_1) + y_1$$
$$= \left(\frac{2 - 0}{1 - 2}\right)(x - 1) + 2$$
$$= -2(x - 1) + 2$$
$$= -2x + 4$$

Piecewise definition

$$f(x) = \begin{cases} \text{line 1} & -2 < x < -1 \\ \text{line 2} & -1 < x < 1 \\ \text{line 3} & 1 < x < 2 \end{cases}$$

$$f(x) = \begin{cases} 2x + 4 & -2 < x < -1 \\ 2 & -1 < x < 1 \\ -2x + 4 & 1 < x < 2 \end{cases}$$

2.3   Periodic Functions

A periodic function is one that repeats its values in a regular interval. The regular interval is called the *period*. The graph of a periodic function looks like a single pattern is being repeated over and over again.

A function $f(x)$ is said to have a period T if

$$f(x + T) = f(x) \quad \text{for all } x.$$

The converse is also true:

If a function has a period T then

$$f(x + T) = f(x) \quad \text{for all } x.$$

The most common examples of periodic functions are the sine and cosine functions. See Fig. 2.7 and Fig. 2.8. These functions repeat themselves every 2π units, so both are periodic with period 2π.

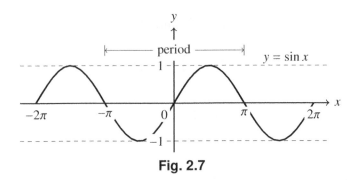

Fig. 2.7

Since $f(x)$ is periodic with period 2π then $f(x + 2\pi) = f(x)$. This is true because $\sin(x + 2\pi) = \sin(x)$

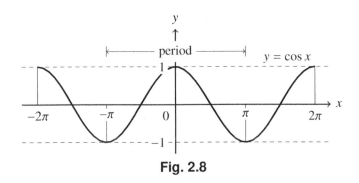

Fig. 2.8

Since $f(x)$ is periodic with period 2π then $f(x + 2\pi) = f(x)$. This is true because $\cos(x + 2\pi) = \cos(x)$

2.4 Periodic Piecewise Functions

These are periodic functions that are built from piecewise functions. If we take a piecewise function $f(x)$ defined in a finite section of the x-axis and repeat it throughout the x-axis, we get a periodic piecewise function.

Consider the piecewise function:

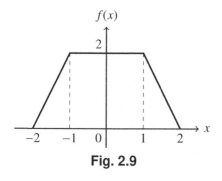

$$f(x) = \begin{cases} 2x + 4 & -2 < x < -1 \\ 2 & -1 < x < 1 \\ -2x + 4 & 1 < x < 2 \end{cases}$$

Fig. 2.9

Now if we repeat the previous function throughout the x-axis we get a periodic piecewise function:

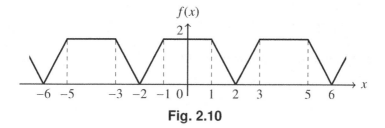

Fig. 2.10

The period of this function is the length of the section of x-axis on which the previous function was defined. Thus

$$T = 2 - (-2)$$
$$= 4$$

Thus $f(x + 4) = f(x)$. But what algebraic expression do we use to define this function? We use the pair of equations below:

$$f(x) = \begin{cases} 2x + 4 & -2 < x < -1 \\ 2 & -1 < x < 1 \\ -2x + 4 & 1 < x < 2 \end{cases} \qquad (2.1)$$

$$f(x + 4) = f(x) \qquad (2.2)$$

This pair of equations is known as the *analytical definition* of the function. Equation (2.1) defines the function between $x = -2$ and $x = 2$, while (2.2) specifies the period of the function.

Study the examples below for a better understanding.

Example 1

Consider the piecewise function below.

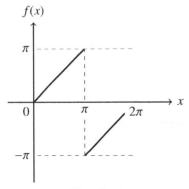

$$f(x) = \begin{cases} x & 0 < x < \pi \\ x - 2\pi & \pi < x < 2\pi \end{cases}$$

Fig. 2.11

If we repeat this function throughout the x-axis we get a periodic piecewise function:

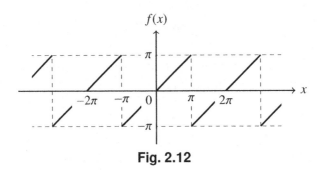

Fig. 2.12

The periodic piecewise function has period 2π. Thus the analytical definition is

$$f(x) = \begin{cases} x & 0 < x < \pi \\ x - 2\pi & \pi < x < 2\pi \end{cases}$$

$$f(x + 2\pi) = f(x)$$

If we take a careful look at Fig. 2.12, we find that the straight line $\nearrow$ going from $(-\pi, -\pi)$ to (π, π) is repeated throughout the graph of $f(x)$. Thus a simpler analytical definition for $f(x)$ is:

$$f(x) = x \qquad -\pi < x < \pi$$

$$f(x + 2\pi) = f(x)$$

Example 2

Find the analytical definition of the function below.

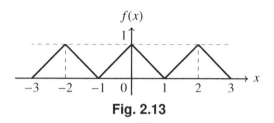

Fig. 2.13

Solution

The function can be observed to repeat its values every 2 units. Thus $f(x)$ has period of 2. Therefore

$$f(x + 2) = f(x)$$

Let us consider one period between $x = -1$ and $x = 1$. Thus a period is made up of two slanted lines: Line 1 and Line 2.

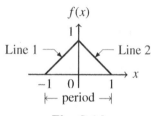

Fig. 2.14

Line 1   Line 1 passes through the points: $(-1, 0)$ and $(0, 1)$. Therefore

$$y = \left(\frac{y_1 - y_2}{x_1 - x_2}\right)(x - x_1) + y_1$$

$$= \left(\frac{0 - 1}{-1 - 0}\right)[x - (-1)] + 0$$

$$= \frac{-1}{-1}(x + 1) + 0$$

$$= x + 1$$

Line 2   Line 2 passes through the points: $(0, 1)$ and $(1, 0)$. Therefore

$$y = \left(\frac{y_1 - y_2}{x_1 - x_2}\right)(x - x_1) + y_1$$

$$= \left(\frac{1 - 0}{0 - 1}\right)(x - 0) + 1$$

$$= \frac{1}{-1}x + 1$$

$$= -x + 1$$

Analytical definition   The analytical definition is

$$f(x) = \begin{cases} x + 1 & -1 < x < 0 \\ -x + 1 & 0 < x < 1 \end{cases}$$

$$f(x + 2) = f(x)$$

Example 3

Draw the graph of the following function.

$$f(x) = \begin{cases} -5 & -1 < x < 0 \\ 5 & 0 < x < 1 \end{cases}$$

$$f(x + 2) = f(x)$$

Solution

The first step is to draw the graph of the piecewise function.   See Fig. 2.15.

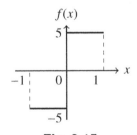

Fig. 2.15

The next and last step is to repeat this graph throughout the x-axis, both left and right. This is shown in Fig. 2.16.

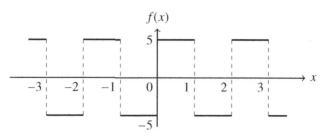

Fig. 2.16

Exercise

1. Find the analytical definition of the function below.

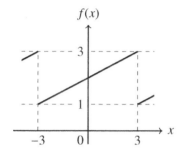

2. Find the analytical definition of the function below.

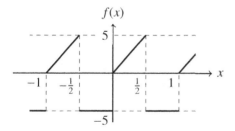

3. Draw the graph of $f(x)$ given the analytical definition below.

$$f(x) = \begin{cases} -1 & -\pi < x < -\dfrac{\pi}{2} \\ \dfrac{2x}{\pi} & -\dfrac{\pi}{2} < x < \dfrac{\pi}{2} \\ 1 & \dfrac{\pi}{2} < x < \pi \end{cases}$$

$$f(x + 2\pi) = f(x)$$

CHAPTER 3

Int-Vectors and Fourier Series

What are int-vectors? Int-vectors make up the tool we would use in place of calculus to solve Fourier series problems. This chapter introduces the idea of int-vectors and, later on, discusses how they are used in Fourier series.

An int-vector is simply an ordered list of two real numbers a and b, written as (a, b), which obey some laws. These laws are given below.

3.1 General Laws of Int-Vectors

Rule 1 $\quad (a, b) = (a, 0) + (0, b)$

Rule 2 $\quad (a, b) = -(b, a)$

We can apply the above laws to simplify expressions made up of int-vectors. The following examples illustrate this.

Example 1

1. $(2, 6) = (2, 0) + (0, 6) \qquad \because$ Rule 1

2. $(-3, 0) + (0, 1) = (-3, 1) \qquad \because$ Rule 1

3. $(0, 2\pi) = -(2\pi, 0) \qquad \because$ Rule 2

4. $(\pi, 2) + (\pi, 2) = 2(\pi, 2) \qquad \because$ basic algebra

5. $5 \times (3, -2) = 5(3, -2) \qquad \because$ basic algebra

6. $5(1, 0) + 5(0, 2) = 5[(1, 0) + (0, 2)] = 5(1, 2) \qquad \because$ Rule 2

Example 2

Simplify the following expressions using Rules 1 and 2 and basic algebra laws.

1. $(2, 3) + (3, 2)$

2. $(3, 3)$

3. $(1, 2) + (1, 3) - (2, 1) + (3, 1)$

4. $(3, 8) - (8, 0) + (0, 3)$

5. $(-1, 7) + (6, 0) + (7, -1) + (0, 6)$

6. $(-\pi, \pi) + (\pi, 0) + (0, -\pi)$

Answers   Here are the answers to Example 2 above. Go through them carefully.

1.

$$(2, 3) + (3, 2) = (2, 3) - (2, 3) \qquad \because \text{Rule 2}$$
$$= 0$$

2.

$$(3, 3) = -(3, 3) \qquad \because \text{Rule 2}$$

add $(3, 3)$ to both sides

$$(3, 3) + (3, 3) = -(3, 3) + (3, 3)$$
$$2(3, 3) = 0$$

divide both sides by 2

$$(3, 3) = \frac{0}{2}$$
$$(3, 3) = 0$$

3.

$$(1, 2) + (1, 3) - (2, 1) + (3, 1)$$

$= (1, 2) - (2, 1) + (1, 3) + (3, 1) \qquad \because \text{rearrange the terms}$

$= (1, 2) - [-(1, 2)] + (1, 3) + [-(1, 3)] \qquad \because \text{Rule 2}$

$= (1, 2) + (1, 2) + (1, 3) - (1, 3)$

$= 2(1, 2) + 0$

$= 2(1, 2)$

4.

$(3, 8) - (8, 0) + (0, 3)$

$= [(3, 0) + (0, 8)] - (8, 0) + (0, 3) \qquad \because \text{Rule 1}$

$= (3, 0) + (0, 8) - (8, 0) + (0, 3)$

$= (3, 0) + (0, 3) + (0, 8) - (8, 0) \qquad \because \text{rearrange the terms}$

$= (3, 0) - (3, 0) + (0, 8) - [-(0, 8)] \qquad \because \text{Rule 2}$

$= 0 + (0, 8) + (0, 8)$

$= 2(0, 8)$

5.

$(-1, 7) + (6, 0) + (7, -1) + (0, 6)$

$= (-1, 7) + (7, -1) + (6, 0) + (0, 6) \qquad \because \text{rearrange the terms}$

$= (-1, 7) - (-1, 7) + (6, 0) - (6, 0) \qquad \because \text{Rule 2}$

$= 0 + 0$

$= 0$

6.

$(-\pi, \pi) + (\pi, 0) + (0, -\pi)$

$= [(-\pi, 0) + (0, \pi)] + (\pi, 0) + (0, -\pi) \qquad \because \text{Rule 1}$

$= (-\pi, 0) + (0, \pi) + (\pi, 0) + (0, -\pi)$

$= (-\pi, 0) + (0, -\pi) + (0, \pi) + (\pi, 0) \qquad \because \text{rearrange the terms}$

$= (-\pi, 0) - (-\pi, 0) + (0, \pi) - (0, \pi) \qquad \because \text{Rule 2}$

$= 0 + 0$

$= 0$

3.2 Int-Vectors with Function Coefficient

Up until now, we have only considered int-vectors with constant coefficients. Int-vectors may also have a *function coefficient*. A *function coefficient*, as the name implies, is a coefficient that is to be interpreted as a function rather than a variable. Function coefficients are represented by letters, e.g., x, y, etc.

For example, in the expression $x(2, 3)$, the coefficient x does not represent a variable but it represents the function $f(x) = x$.

Example

Simplify the following expressions using Rules 1 and 2 and basic algebra laws.

1. $x(6, 0) - (0, 6)(x + 1)$

2. $x(3, 2) + (x + \pi)(2, 3)$

Solution

$x(6, 0) - (0, 6)(x + 1)$

$= x(6, 0) + (6, 0)(x + 1)$

factorise out (6,0)

$= (6, 0)(x + x + 1)$

$= (6, 0)(2x + 1)$

$= (2x + 1)(6, 0)$

Solution

$x(3, 2) + (x + \pi)(2, 3)$

$= x(3, 2) - (x + \pi)(3, 2)$

factorise out (3,2)

$= (3, 2)(x - x - \pi)$

$= (3, 2)(0 - \pi)$

$= (3, 2)(-\pi)$

$= -\pi(3, 2)$

3.3 Fourier Series

The Fourier series of a function $f(x)$ with period 2π is

$$f(x) = \frac{a_0}{2} + \sum_{n=1}^{\infty} a_n \cos nx + \sum_{n=1}^{\infty} b_n \sin nx \qquad (3.1)$$

The constants a_0, a_n, b_n are called *Fourier coefficients*. In any Fourier series problem most of the work done is solving for these constants. How do we solve for the Fourier coefficients? This is the main discussion of the next section.

3.4   Fourier Coefficients

Fourier coefficients are the constants a_0, a_n, and b_n. Here a_n represents constants a_1, a_2, a_3, ..., etc. Similarly, b_n represents constants b_1, b_2, b_3, ..., etc. We need to solve for the values of the Fourier coefficients before we can get the Fourier series of a function. We employ int-vectors when solving for a_n and b_n.

The int-vectors have different values depending on whether we are solving for a_n or b_n.

Rules when solving for a_n

Rule 3a $(0, a) = \dfrac{1}{n} \sin na$

Rule 4a $x(0, a) = \dfrac{a}{n} \sin na + \dfrac{1}{n^2}(\cos na - 1)$

Rules when solving for b_n

Rule 3b $(0, a) = \dfrac{1}{n}(1 - \cos na)$

Rule 4b $x(0, a) = -\dfrac{a}{n} \cos na + \dfrac{1}{n^2} \sin na$

Where did the n come from? You may be wondering about the n that appears in the formulas above. The n is a positive integer variable such that $n = 1$, 2, 3, ..., etc. It is same as the subscript of a_n and b_n.

Now make a note of the formulas above. You will need them when solving for Fourier series problems.

Example 1

Assume you are *solving for a_n*. Use Rules 3a and 4a to evaluate the following expressions.

1. $(0, 6)$

2. $(\pi, 0)$

3. $x(0, 3)$

4. $x(0, 2\pi)$

Solution

1.

$$(0, 6) = \frac{1}{n} \sin 6n$$

2.

$$(\pi, 0) = -(0, \pi)$$

$$= -[\frac{1}{n} \sin \pi n]$$

$$= \frac{-1}{n} \sin \pi n$$

$$= \frac{-1}{n} \times 0$$

$$= 0$$

4.

$$x(0, 2\pi) = \frac{2\pi}{n} \sin n(2\pi)$$

$$+ \frac{1}{n^2}(\cos n(2\pi) - 1)$$

$$= \frac{2\pi}{n} \sin 2\pi n$$

$$+ \frac{1}{n^2}(\cos 2\pi n - 1)$$

$$= \frac{2\pi}{n} \times 0 + \frac{1}{n^2}(1 - 1)$$

$$= 0 + 0$$

$$= 0$$

3.

$$x(0, 3) = \frac{3}{n} \sin n(3) + \frac{1}{n^2}(\cos n(3) - 1)$$

$$= \frac{3}{n} \sin 3n + \frac{1}{n^2}(\cos 3n - 1)$$

Example 2

Assume you are *solving for b_n*. Use Rules 3b and 4b to evaluate the following expressions.

1. $(0, 6)$

2. $(\pi, 0)$

3. $x(0, 3)$

4. $x(0, 2\pi)$

Solution

1.

$$(0, 6) = \frac{1}{n}(1 - \cos 6n)$$

2.

$$(\pi, 0) = -(0, \pi)$$

$$= -\frac{1}{n}[1 - \cos n(\pi)]$$

$$= \frac{-1}{n}(1 - \cos \pi n)$$

$$= \frac{1}{n}(-1 + \cos \pi n)$$

$$= \frac{1}{n}(\cos \pi n - 1)$$

3.

$$x(0, 3) = -\frac{3}{n}\cos n(3) + \frac{1}{n^2}\sin n(3)$$

$$= -\frac{3}{n}\cos 3n + \frac{1}{n^2}\sin 3n$$

4.

$$x(0, 2\pi) = -\frac{2\pi}{n}\cos n(2\pi)$$

$$+ \frac{1}{n^2}\sin n(2\pi)$$

$$= -\frac{2\pi}{n}\cos 2\pi n + \frac{1}{n^2}\sin 2\pi n$$

$$= -\frac{2\pi}{n} \times 1 + \frac{1}{n^2} \times 0$$

$$= -\frac{2\pi}{n} + 0$$

$$= -\frac{2\pi}{n}$$

3.5 Fourier Series Problems

Now that we have some knowledge of Fourier coefficients, we are ready to start solving Fourier series problems. Study the following example carefully. The routine for solving problems is given afterwards.

Example 1

Find the Fourier series of the function $f(x)$ defined thus:

$$f(x) = \begin{cases} -1 & -\pi < x < 0 \\ 1 & 0 < x < \pi \end{cases}$$

$$f(x + 2\pi) = f(x)$$

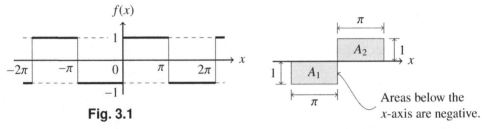

f(x)
1
−1
x
−2π
−π
0
π
2π

Fig. 3.1

π
A2
1
x
1
A1
π
Areas below the
x-axis are negative.

Fig. 3.2

Solution for graph of $f(x)$ First, we need to sketch the graph of the function $f(x)$. This is shown in Fig. 3.1.

Next, we shall solve for a_0, a_n and b_n.

Solution for a_0 The formula for a_0 is:

$$a_0 = \frac{A}{\pi}$$

where A is the net area of $f(x)$ in one period, i.e, from $x = -\pi$ to $x = \pi$. This is shown in Fig. 3.2.

$$\begin{aligned} A &= -A_1 + A_2 \\ &= -(\pi \times 1) + (\pi \times 1) \\ &= -\pi + \pi \\ &= 0 \end{aligned}$$

Therefore

$$a_0 = \frac{A}{\pi} = \frac{0}{\pi}$$

$$\boxed{a_0 = 0}$$

Solution for a_n Recall that

$$f(x) = \begin{cases} -1 & -\pi < x < 0 \\ 1 & 0 < x < \pi \end{cases}$$

The definition of the function is made up of two cases. We can write the cases as:

$$-1 \quad (-\pi, 0)$$
$$1 \quad (0, \pi)$$

The Fourier coefficient a_n multiplied by π is the *sum* of the two cases. Thus

$$a_n \pi = -1(-\pi, 0) + 1(0, \pi)$$

We can simplify the above equation using the Rules we have learned so far.

$$a_n \pi = (0, -\pi) + (0, \pi)$$

apply Rule 3a

$$= \frac{1}{n} \sin n(-\pi) + \frac{1}{n} \sin n(\pi)$$
$$= 0 + 0$$
$$a_n \pi = 0$$
Therefore
$$\boxed{a_n = 0}$$

Solution for b_n Recall that

$$f(x) = \begin{cases} -1 & -\pi < x < 0 \\ 1 & 0 < x < \pi \end{cases}$$

The definition of the function is made up of two cases. We can write the cases as:

$$-1 \quad (-\pi, 0)$$
$$1 \quad (0, \pi)$$

The Fourier coefficient b_n multiplied by π is the *sum* of the two cases. Thus

$$b_n \pi = -1(-\pi, 0) + 1(0, \pi)$$

We can simplify the above equation using the Rules we have learned so far.

$$b_n \pi = (0, -\pi) + (0, \pi)$$

apply Rule 3b

$$= \frac{1}{n}[1 - \cos n(-\pi)] + \frac{1}{n}[1 - \cos n(\pi)]$$

$$= \frac{1}{n}(1 - \cos n\pi) + \frac{1}{n}(1 - \cos n\pi)$$

$$= \frac{2}{n}(1 - \cos n\pi)$$

Thus

$$b_n = \frac{2}{n\pi}(1 - \cos n\pi)$$

Recall that

$$\cos n\pi = \begin{cases} 1 & n \text{ is even} \\ -1 & n \text{ is odd} \end{cases}$$

Therefore

$$b_n = \begin{cases} 0 & n \text{ is even} \\ \dfrac{4}{n\pi} & n \text{ is odd} \end{cases}$$

Solution for Fourier Series

$$f(x) = \frac{a_0}{2} + \sum_{n=1}^{\infty} a_n \cos nx + \sum_{n=1}^{\infty} b_n \sin nx$$

Put $a_0 = 0$ and $a_n = 0$

$$f(x) = 0 + 0 + \sum_{n=1}^{\infty} b_n \sin nx$$

$$= b_n(\sin x + \sin 2x + \sin 3x + \sin 4x + \cdots)$$

separate the odd and even terms

$$= b_n(\sin x + \sin 3x + \sin 5x + \cdots)$$
$$+ b_n(\sin 2x + \sin 4x + \sin 6x + \cdots)$$

$$\text{put } b_n = 0 \text{ for even } n, \ b_n = \frac{4}{n\pi} \text{ for odd } n$$

$$= \frac{4}{n\pi}(\sin x + \sin 3x + \sin 5x + \cdots)$$
$$+ 0(\sin 2x + \sin 4x + \sin 6x + \cdots)$$
$$= \frac{4}{\pi}(\sin x + \frac{1}{3} \sin 3x + \frac{1}{5} \sin 5x + \cdots)$$

Therefore

$$\boxed{f(x) = \frac{4}{\pi}(\sin x + \frac{1}{3} \sin 3x + \frac{1}{5} \sin 5x + \cdots)}$$

This is the Fourier series expansion of $f(x)$.

Observation from Example 1

Go through the working of Example 1 carefully. You will notice that the only difference between the solutions for a_n and b_n is the rule applied. The Fourier coefficient a_n used Rule 3a while b_n used Rule 3b. Therefore we can shorten our working by reusing the result we get for $a_n\pi$, just before we apply Rule 3a, to solve for b_n.

Shorter Solution of a_n and b_n

Solution for a_n   Recall that

$$f(x) = \begin{cases} -1 & -\pi < x < 0 \\ 1 & 0 < x < \pi \end{cases}$$

The definition of the function is made up of two cases. We can write the cases as:

$$-1 \quad (-\pi, 0)$$
$$1 \quad (0, \pi)$$

The Fourier coefficient a_n multiplied by π is the *sum* of the two cases. Thus

$$a_n\pi = -1(-\pi, 0) + 1(0, \pi)$$

We can simplify the above equation using the Rules we have learned so far.

$$a_n \pi = -1(-\pi, 0) + 1(0, \pi)$$
$$= 1(0, -\pi) + 1(0, \pi)$$
$$= (0, -\pi) + (0, \pi) \qquad \text{(reuse point)}$$

apply Rule 3a

$$= \frac{1}{n} \sin n(-\pi) + \frac{1}{n} \sin n(\pi)$$
$$= -\frac{1}{n} \sin n\pi + \frac{1}{n} \sin n\pi$$
$$= -0 + 0$$
$$= 0$$

Therefore

$$\boxed{a_n = 0}$$

Solution for b_n   From the reuse point,

$$a_n \pi = (0, -\pi) + (0, \pi)$$

Therefore

$$b_n \pi = (0, -\pi) + (0, \pi)$$

apply Rule 3b

$$= \frac{1}{n}[1 - \cos n(-\pi)] + \frac{1}{n}[1 - \cos n(\pi)]$$
$$= \frac{1}{n}(1 - \cos n\pi) + \frac{1}{n}(1 - \cos n\pi)$$
$$= \frac{2}{n}(1 - \cos n\pi)$$

Therefore

$$b_n = \frac{2}{n\pi}(1 - \cos n\pi)$$

Recall that

$$\cos n\pi = \begin{cases} 1 & n \text{ is even} \\ -1 & n \text{ is odd} \end{cases}$$

Therefore

$$b_n = \begin{cases} 0 & n \text{ is even} \\ \dfrac{4}{n\pi} & n \text{ is odd} \end{cases}$$

3.5.1 Routine for Solving Problems

1. Prerequisites: If you are given the graph of the function, first find the analytical definition of the function. If you are given the analytical definition of the function, first draw the graph of the function.

2. Solve for a_0: Determine the area (A) of one period of the function. The Fourier coefficient a_0 is given by the formula:

$$a_0 = \frac{A}{\pi}$$

3. Solve for a_n: The formula for a_n is given as:

$$a_n \pi = \text{sum of cases}$$

Apply only the general rules to the int-vectors until all int-vectors are of the form $(0, a)$ or $x(0, a)$ for some real number a. This point in the working is labeled as the *reuse point*. Then apply Rule 3a and/or Rule 4a.

4. Solve for b_n: Solve for b_n by continuing from the reuse point. Apply Rule 3b and/or Rule 4b.

5. Solve for Fourier series: The Fourier series for a function $f(x)$ is

$$f(x) = \frac{a_0}{2} + \sum_{n=1}^{\infty} a_n \cos nx + \sum_{n=1}^{\infty} b_n \sin nx$$

Simply substitute a_0, a_n and b_n with their respective values according to what has been solved.

Example 2

Find the Fourier series for the function shown in Fig. 3.3. Assume it has period 2π.

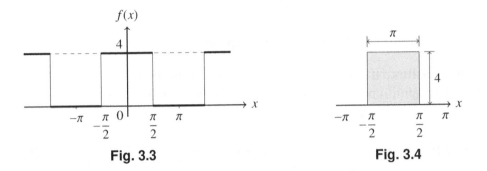

Fig. 3.3   **Fig. 3.4**

Solution for analytical definition   First, we would define the function analytically.

$$f(x) = \begin{cases} 0 & -\pi < x < -\dfrac{\pi}{2} \\[2mm] 4 & -\dfrac{\pi}{2} < x < \dfrac{\pi}{2} \\[2mm] 0 & \dfrac{\pi}{2} < x < \pi \end{cases}$$

$$f(x + 2\pi) = f(x)$$

Solution of a_0   The area covered by the graph in one period ($x = -\pi$ to $x = \pi$) is shown in Fig. 3.4.

Notice that in the intervals $-\pi < x < -\frac{\pi}{2}$ and $\frac{\pi}{2} < x < \pi$, the lines have zero height above the x-axis. Thus the area in these intervals in zero.

From Fig. 3.4 the area A in one period is calculated as

$$A = 4 \times \pi$$
$$= 4\pi$$

Thus

$$a_0 = \frac{A}{\pi}$$
$$= \frac{4\pi}{\pi}$$

$$\boxed{a_0 = 4}$$

Solution of a_n   The function $f(x)$ is defined in one period as:

$$f(x) = \begin{cases} 0 & -\pi < x < -\dfrac{\pi}{2} \\ 4 & -\dfrac{\pi}{2} < x < \dfrac{\pi}{2} \\ 0 & \dfrac{\pi}{2} < x < \pi \end{cases}$$

The definition of the function is made up of three cases. We can write the cases as:

$$0 \quad (-\pi, -\frac{\pi}{2})$$

$$4 \quad (-\frac{\pi}{2}, \frac{\pi}{2})$$

$$0 \quad (\frac{\pi}{2}, \pi)$$

The Fourier coefficient a_n multiplied by π is the sum of the cases.

$$a_n\pi = 0(-\pi, -\frac{\pi}{2}) + 4(-\frac{\pi}{2}, \frac{\pi}{2}) + 0(\frac{\pi}{2}, \pi)$$

$$= 0 + 4(-\frac{\pi}{2}, \frac{\pi}{2}) + 0$$

$$= 4(-\frac{\pi}{2}, \frac{\pi}{2})$$

$$= 4(-\frac{\pi}{2}, 0) + 4(0, \frac{\pi}{2})$$

$$= -4(0, -\frac{\pi}{2}) + 4(0, \frac{\pi}{2})$$

$$\frac{a_n\pi}{4} = -(0, -\frac{\pi}{2}) + (0, \frac{\pi}{2}) \qquad \text{(reuse point)}$$

apply Rule 3a

$$= -\frac{1}{n}\sin n(-\frac{\pi}{2}) + \frac{1}{n}\sin n(\frac{\pi}{2})$$

$$= \frac{1}{n}\sin \frac{n\pi}{2} + \frac{1}{n}\sin \frac{n\pi}{2}$$

$$\frac{a_n\pi}{4} = \frac{2}{n}\sin \frac{n\pi}{2}$$

$$a_n = \frac{8}{n\pi}\sin \frac{n\pi}{2}$$

Recall that

$$\sin \frac{n\pi}{2} = \begin{cases} 0 & n \text{ is even} \\ 1 & n = 1, 5, 9, \ldots \\ -1 & n = 3, 7, 11, \ldots \end{cases}$$

Therefore

$$a_n = \begin{cases} 0 & n \text{ is even} \\ \dfrac{8}{n\pi} & n = 1, 5, 9, \ldots \\ -\dfrac{8}{n\pi} & n = 3, 7, 11, \ldots \end{cases}$$

Solution for b_n   Continuing from the reuse point,

$$\frac{b_n\pi}{4} = -(0, -\frac{\pi}{2}) + (0, \frac{\pi}{2})$$

apply Rule 3b

$$= -\frac{1}{n}[1 - \cos n(-\frac{\pi}{2})] + \frac{1}{n}[1 - \cos n(\frac{\pi}{2})]$$

$$= -\frac{1}{n}(1 - \cos \frac{n\pi}{2}) + \frac{1}{n}(1 - \cos \frac{n\pi}{2})$$

$$\boxed{b_n = 0}$$

Solution for Fourier series

$$f(x) = \frac{a_0}{2} + \sum_{n=1}^{\infty} a_n \cos nx + \sum_{n=1}^{\infty} b_n \sin nx$$

$$= \frac{4}{2} + a_n(\cos x + \cos 2x + \cos 3x + \cdots)$$

$$+ b_n(\sin x + \sin 2x + \sin 3x + \cdots)$$

put $b_n = 0$ and separate the cosine terms

$$= 2 + a_n(\cos 2x + \cos 4x + \cdots) + a_n(\cos x + \cos 5x + \cdots)$$

$$+ a_n(\cos 3x + \cos 7x + \cdots) + 0$$

$$= 2 + 0 + \frac{8}{n\pi}(\cos x + \cos 5x + \cdots) - \frac{8}{n\pi}(\cos 3x + \cos 7x + \cdots)$$

$$= 2 + \frac{8}{\pi}(\frac{1}{1}\cos x + \frac{1}{5}\cos 5x + \cdots) - \frac{8}{\pi}(\frac{1}{3}\cos 3x + \frac{1}{7}\cos 7x + \cdots)$$

$$= 2 + \frac{8}{\pi}(\cos x + \frac{1}{5}\cos 5x + \cdots) - \frac{8}{\pi}(\frac{1}{3}\cos 3x + \frac{1}{7}\cos 7x + \cdots)$$

$$\boxed{f(x) = 2 + \frac{8}{\pi}(\cos x - \frac{1}{3}\cos 3x + \frac{1}{5}\cos 5x - \frac{1}{7}\cos 7x + \cdots)}$$

Example 3

Find the Fourier series for the function defined by

$$f(x) = \begin{cases} -x & -\pi < x < 0 \\ 0 & 0 < x < \pi \end{cases}$$

$$f(x + 2\pi) = f(x)$$

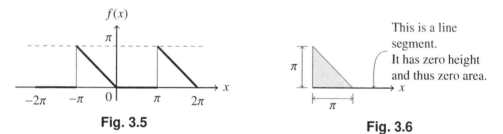

Fig. 3.5   Fig. 3.6

Solution for graph of $f(x)$ We have drawn the graph of the function in Fig. 3.5.

Solution for a_0 The area covered by the graph in one period ($x = -\pi$ to $x = \pi$) is shown in Fig. 3.6.

$$A = \frac{1}{2}\text{base} \times \text{height}$$

$$= \frac{1}{2}(\pi) \times \pi$$

$$= \frac{\pi^2}{2}$$

Thus

$$a_0 = \frac{A}{\pi}$$

$$= \frac{\pi^2}{2\pi}$$

$$\boxed{a_0 = \frac{\pi}{2}}$$

Solution for a_n

$$a_n\pi = \text{sum of cases}$$
$$= -x(-\pi, 0) + 0(0, \pi)$$
$$= -x(-\pi, 0)$$
$$= x(0, -\pi) \qquad \text{(reuse point)}$$

apply Rule 4a

$$= \frac{-\pi}{n} \sin n(-\pi) + \frac{1}{n^2}[\cos n(-\pi) - 1]$$

$$= -\frac{\pi}{n}(0) + \frac{1}{n^2}(\cos n\pi - 1)$$

$$a_n\pi = \frac{1}{n^2}(\cos n\pi - 1)$$

$$a_n = \frac{1}{\pi n^2}(\cos n\pi - 1)$$

Therefore

$$\boxed{a_n = \begin{cases} -\dfrac{2}{\pi n^2} & n \text{ is odd} \\ 0 & n \text{ is even} \end{cases}}$$

Solution for b_n Continuing from the reuse point,

$$b_n\pi = x(0, -\pi)$$

apply Rule 4b

$$= -\frac{(-\pi)}{n} \cos n(-\pi) + \frac{1}{n^2} \sin n(-\pi)$$

$$= \frac{\pi}{n} \cos n\pi + 0$$

$$b_n \pi = \frac{\pi}{n} \cos n\pi$$

$$b_n = \frac{1}{n} \cos n\pi$$

Therefore

$$b_n = \begin{cases} -\frac{1}{n} & n \text{ is odd} \\ \frac{1}{n} & n \text{ is even} \end{cases}$$

Solution for Fourier Series

$$= \frac{a_0}{2} + \sum_{n=1}^{\infty} a_n \cos nx + \sum_{n=1}^{\infty} b_n \sin nx$$

$$= \frac{\pi/2}{2} + a_n(\cos x + \cos 2x + \cos 3x + \cdots)$$
$$+ b_n(\sin x + \sin 2x + \sin 3x + \cdots)$$

separate odd and even terms

$$= \frac{\pi}{4} + a_n(\cos x + \cos 3x + \cdots) + a_n(\cos 2x + \cos 4x + \cdots)$$
$$+ b_n(\sin x + \sin 3x + \cdots) + b_n(\sin 2x + \sin 4x + \cdots)$$

$$= \frac{\pi}{4} - \frac{2}{\pi n^2}(\cos x + \cos 3x + \cdots) + 0(\cos 2x + \cos 4x + \cdots)$$
$$- \frac{1}{n}(\sin x + \sin 3x + \cdots) + \frac{1}{n}(\sin 2x + \sin 4x + \cdots)$$

$$= \frac{\pi}{4} - \frac{2}{\pi}(\frac{1}{1^2} \cos x + \frac{1}{3^2} \cos 3x + \cdots)$$
$$- (\frac{1}{1} \sin x + \frac{1}{3} \sin 3x + \cdots) + (\frac{1}{2} \sin 2x + \frac{1}{4} \sin 4x + \cdots)$$

$$= \frac{\pi}{4} - \frac{2}{\pi}(\cos x + \frac{1}{3^2} \cos 3x + \cdots)$$
$$- (\sin x - \frac{1}{2} \sin 2x + \frac{1}{3} \sin 3x - \frac{1}{4} \sin 4x + \cdots)$$

$$\boxed{f(x) = \frac{\pi}{4} - \frac{2}{\pi}(\cos x + \frac{1}{3^2} \cos 3x + \cdots) - (\sin x - \frac{1}{2} \sin 2x + \cdots)}$$

Example 4

Determine the Fourier series to represent the periodic function shown in Fig. 3.7.

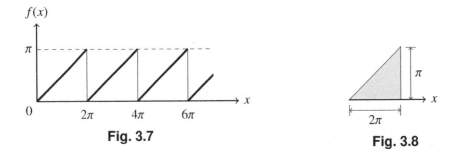

Fig. 3.7   Fig. 3.8

Solution for analytical definition   First we would define the function analytically;

$$f(x) = \frac{x}{2} \quad 0 < x < 2\pi$$

$$f(x + 2\pi) = f(x)$$

Solution for a_0   The area under the curve between $x = 0$ and $x = 2\pi$ is that of a *triangle* as shown in Fig. 3.8.
The area A of triangle is:

$$A = \frac{1}{2}\text{base} \times \text{height}$$

$$= \frac{1}{2}(2\pi) \times \pi$$

$$= \pi \times \pi$$

$$= \pi^2$$

Thus

$$a_0 = \frac{A}{\pi}$$

$$= \frac{\pi^2}{\pi}$$

$$\boxed{a_0 = \pi}$$

Solution for a_n

$$a_n \pi = \text{sum of cases}$$

$$a_n \pi = \frac{x}{2}(0, 2\pi)$$

$$a_n(2\pi) = x(0, 2\pi) \qquad \text{(reuse point)}$$

apply Rule 4a

$$= \frac{2\pi}{n} \sin n(2\pi) + \frac{1}{n^2}[\cos n(2\pi) - 1]$$

$$= 0 + \frac{1}{n^2}(\cos 2\pi n - 1)$$

$$= \frac{1}{n^2}(1 - 1) = 0$$

$$a_n(2\pi) = 0$$

$$\boxed{a_n = 0}$$

Solution for b_n Continuing from the reuse point,

$$b_n(2\pi) = x(0, 2\pi)$$

apply Rule 4b

$$= -\frac{2\pi}{n} \cos n(2\pi) + \frac{1}{n^2} \sin n(2\pi)$$

$$= -\frac{2\pi}{n}(1) + 0$$

$$b_n(2\pi) = -\frac{2\pi}{n}$$

$$\boxed{b_n = -\frac{1}{n}}$$

Solution for Fourier series

$$f(x) = \frac{a_0}{2} + \sum_{n=1}^{\infty} a_n \cos nx + \sum_{n=1}^{\infty} b_n \sin nx$$

$$= \frac{\pi}{2} + 0 + \sum_{n=1}^{\infty} b_n \sin nx$$

$$= \frac{\pi}{2} + b_n(\sin x + \sin 2x + \cdots)$$

$$= \frac{\pi}{2} - \frac{1}{n}(\sin x + \sin 2x + \cdots)$$

$$\boxed{f(x) = \frac{\pi}{2} - (\sin x + \frac{1}{2}\sin 2x + \frac{1}{3}\sin 3x + \cdots)}$$

Exercise 1

Determine the Fourier series for periodic functions whose graphs are shown below. Assume they are all periodic with period 2π.

1.

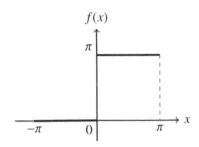

3.

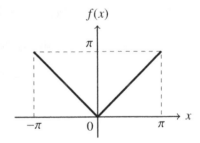

2.

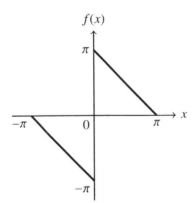

4.

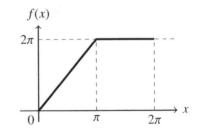

Exercise 2

Determine the Fourier series for the following periodic functions. Assume they are all periodic with period 2π.

1.

$$f(x) = x + \pi \qquad -\pi < x < \pi$$

2.

$$f(x) = \begin{cases} 0 & -\pi < x < 0 \\ 1 & 0 < x < \pi \end{cases}$$

3.

$$f(x) = \begin{cases} -2 & -\pi < x < 0 \\ 2 & 0 < x < \pi \end{cases}$$

4.

$$f(x) = \begin{cases} 1 + \dfrac{2x}{\pi} & -\pi < x < 0 \\ 1 - \dfrac{2x}{\pi} & 0 < x < \pi \end{cases}$$

5.

$$f(x) = \begin{cases} 0 & -\pi < x < 0 \\ 1 & 0 < x < \dfrac{\pi}{2} \\ -1 & \dfrac{\pi}{2} < x < \pi \end{cases}$$

6.

$$f(x) = \begin{cases} 0 & -\pi < x < 0 \\ x & 0 < x < \dfrac{\pi}{2} \\ \pi - x & \dfrac{\pi}{2} < x < \pi \end{cases}$$

CHAPTER 4

Secondary Rules of Int-Vectors

In this chapter, we shall discuss some additional rules which int-vectors obey. These rules are not necessary to find the Fourier series of a function but they make the work easier.

4.1 How to Find a_0 without Drawing the Graph of $f(x)$

In all previous examples, we found a_0 by drawing the graph of $f(x)$ and calculating the area under one period. This section explains how to find a_0 without drawing the graph of $f(x)$.

If we are given the analytical definition of a function we can find a_0 directly by using the formula:

$$a_0\pi = \text{sum of cases}$$

Int-vectors obey the following rules when we are solving for a_0.

Rules when solving for a_0

Rule 3c $(0, a) = a$

Rule 4c $x(0, a) = \dfrac{a^2}{2}$

Example 1

Find a_0 for the function $f(x)$ defined thus:

$$f(x) = \begin{cases} -5 & -\pi < x < 0 \\ 5 & 0 < x < \pi \end{cases}$$

Assume it is periodic with period 2π.

Solution

The function $f(x)$ is made up of two cases.

$$-5 \quad (-\pi, 0)$$
$$5 \quad (0, \pi)$$

$$a_0\pi = \text{sum of cases}$$
$$= -5(-\pi, 0) + 5(0, \pi)$$
$$= 5(0, -\pi) + 5(0, \pi)$$

apply Rule 3c

$$= 5(-\pi) + 5(\pi)$$
$$= -5\pi + 5\pi$$
$$a_0\pi = 0$$
$$\boxed{a_0 = 0}$$

Example 2

Find a_0 for the function:

$$f(x) = \begin{cases} -\pi & -\pi < x < 0 \\ x & 0 < x < \pi \end{cases}$$

Assume it is periodic with period 2π.

Solution

$$a_0\pi = \text{sum of cases}$$
$$= -\pi(-\pi, 0) + x(0, \pi)$$
$$= \pi(0, -\pi) + x(0, \pi)$$

apply Rules 3c and 4c

$$= \pi(-\pi) + \frac{\pi^2}{2}$$
$$= -\pi^2 + \frac{\pi^2}{2}$$
$$a_0\pi = -\frac{\pi^2}{2}$$

$$\boxed{a_0 = -\frac{\pi}{2}}$$

Example 3

Find the Fourier series for the function

$$f(x) = \begin{cases} 1 & -\pi < x < 0 \\ 3 & 0 < x < \pi \end{cases}$$

Assume it is periodic with period 2π.

Solution for a_0

$$a_0\pi = \text{sum of cases}$$
$$= 1(-\pi, 0) + 3(0, \pi)$$
$$= -(0, -\pi) + 3(0, \pi) \qquad \text{(reuse point)}$$

apply Rule 3c

$$= -(-\pi) + 3(\pi)$$
$$= \pi + 3\pi$$
$$a_0\pi = 4\pi$$
$$\boxed{a_0 = 4}$$

Solution for a_n   Continuing from the reuse point, we get:

$$a_n\pi = -(0, -\pi) + 3(0, \pi)$$

apply Rule 3a

$$= -\frac{1}{n}\sin n(-\pi) + \frac{3}{n}\sin n\pi$$
$$= -0 + 0$$
$$a_n\pi = 0$$
$$\boxed{a_n = 0}$$

Solution for b_n Continuing from the reuse point, we get:

$$b_n \pi = -(0, -\pi) + 3(0, \pi)$$

apply Rule 3b

$$= -\frac{1}{n}[1 - \cos n(-\pi)] + \frac{3}{n}[1 - \cos n\pi]$$

$$= \frac{1}{n}(-1 + \cos n\pi + 3 - 3\cos n\pi)$$

$$= \frac{1}{n}(2 - 2\cos n\pi)$$

$$b_n \pi = \frac{2}{n}(1 - \cos n\pi)$$

$$b_n = \frac{2}{n\pi}(1 - \cos n\pi)$$

Therefore

$$b_n = \begin{cases} \dfrac{4}{n\pi} & n \text{ is odd} \\ 0 & n \text{ is even} \end{cases}$$

Solution for Fourier series

$$f(x) = \frac{a_0}{2} + \sum_{n=1}^{\infty} a_n \cos nx + \sum_{n=1}^{\infty} b_n \sin nx$$

$$= \frac{4}{2} + 0 + \sum_{n=1}^{\infty} b_n \sin nx$$

$$= 2 + b_n(\sin x + \sin 2x + \cdots)$$

$$= 2 + \frac{4}{n\pi}(\sin x + \sin 3x + \cdots) + 0(\sin 2x + \sin 4x + \cdots)$$

$$\boxed{f(x) = 2 + \frac{4}{\pi}(\sin x + \frac{1}{3}\sin 3x + \cdots)}$$

Example 4

Find the Fourier series of the function $f(x)$ which has a period of 2π.

$$f(x) = \begin{cases} x & 0 < x < \pi \\ \pi & \pi < x < 2\pi \end{cases}$$

Solution for a_0

$$\begin{aligned} a_0\pi &= x(0, \pi) + \pi(\pi, 2\pi) \\ &= x(0, \pi) + \pi(\pi, 0) + \pi(0, 2\pi) \\ &= x(0, \pi) - \pi(0, \pi) + \pi(0, 2\pi) \qquad \text{(reuse point)} \end{aligned}$$

apply Rules 3c and 4c

$$\begin{aligned} &= \frac{\pi^2}{2} - \pi(\pi) + \pi(2\pi) \\ &= \frac{\pi^2}{2} - \pi^2 + 2\pi^2 \\ a_0\pi &= \frac{3\pi^2}{2} \end{aligned}$$

$$\boxed{a_0 = \frac{3\pi}{2}}$$

Solution for a_n   Continuing from the reuse point, we get:

$$a_n\pi = x(0, \pi) - \pi(0, \pi) + \pi(0, 2\pi)$$

apply Rules 3a and 4a

$$\begin{aligned} &= [\frac{\pi}{n}\sin n\pi + \frac{1}{n^2}(\cos n\pi - 1)] - \frac{\pi}{n}\sin n\pi + \frac{\pi}{n}\sin n(2\pi) \\ &= 0 + \frac{1}{n^2}(\cos n\pi - 1) - 0 + 0 \\ a_n &= \frac{1}{\pi n^2}(\cos n\pi - 1) \end{aligned}$$

Therefore

$$a_n = \begin{cases} -\dfrac{2}{\pi n^2} & n \text{ is odd} \\ 0 & n \text{ is even} \end{cases}$$

Solution for b_n   Continuing from the reuse point, we get:

$$b_n\pi = x(0, \pi) - \pi(0, \pi) + \pi(0, 2\pi)$$

apply Rules 3b and 4b

$$= [-\frac{\pi}{n}\cos n\pi + \frac{1}{n^2}\sin n\pi] - \frac{\pi}{n}(1 - \cos n\pi) + \frac{\pi}{n}(1 - \cos n(2\pi))$$

$$= -\frac{\pi}{n}\cos n\pi + 0 - \frac{\pi}{n}(1 - \cos n\pi) + \frac{\pi}{n}(1 - 1)$$

$$= -\frac{\pi}{n}\cos n\pi - \frac{\pi}{n}(1 - \cos n\pi) + 0$$

$$= \frac{\pi}{n}(-\cos n\pi - 1 + \cos n\pi)$$

$$b_n\pi = \frac{\pi}{n}(-1)$$

$$\boxed{b_n = -\frac{1}{n}}$$

Solution for Fourier series

$$f(x) = \frac{a_0}{2} + \sum_{n=1}^{\infty} a_n \cos nx + \sum_{n=1}^{\infty} b_n \sin nx$$

$$= \frac{3\pi}{4} - \frac{2}{\pi n^2}(\cos x + \cos 3x + \cdots) + 0(\cos 2x + \cos 4x + \cdots)$$

$$- \frac{1}{n}(\sin x + \sin 2x + \cdots)$$

$$\boxed{f(x) = \frac{3\pi}{4} - \frac{2}{\pi}(\cos x + +\frac{1}{3^2}\cos 3x + \cdots) - (\sin x + \frac{1}{2}\sin 2x + \cdots)}$$

Exercise 1

Find the Fourier series for the following functions. Assume they are periodic with period 2π. Solve for a_0 without drawing the graph of $f(x)$.

1.

$$f(x) = 3 - 2x \qquad -\pi < x < \pi$$

2.

$$f(x) = \begin{cases} -1 & -\pi < x < 0 \\ 2 & 0 < x < \pi \end{cases}$$

3.

$$f(x) = \begin{cases} \dfrac{x}{2} & 0 < x < \pi \\ \pi - \dfrac{x}{2} & \pi < x < 2\pi \end{cases}$$

4.

$$f(x) = \begin{cases} -1 & -\pi < x < -\dfrac{\pi}{2} \\ 0 & -\dfrac{\pi}{2} < x < \dfrac{\pi}{2} \\ 1 & \dfrac{\pi}{2} < x < \pi \end{cases}$$

5.

$$f(x) = \begin{cases} -x & -\pi < x < -\dfrac{\pi}{2} \\ 0 & -\dfrac{\pi}{2} < x < \dfrac{\pi}{2} \\ x & \dfrac{\pi}{2} < x < \pi \end{cases}$$

4.2 How to Get Rid of Negative Values

Sometimes in our working, we come across int-vectors containing negative values, e.g., $(0, -\pi)$, $(0, -\frac{\pi}{2})$, etc. It would be nice if we could simplify these int-vectors such that we only deal with positive numbers. There are two major reasons why this simplification is important:

1. It enables us to simplify expressions such as

$$(-3, 0) + (3, 0)$$

As we shall find out in the next set of examples, the above expression can be simplified to either $2(3, 0)$ or zero.

2. Substituting a positive value into a formula is much easier than substituting a negative value.

4.2.1 Negative int-vectors when solving for a_0

The two types of negative int-vectors encountered when solving for a_0 are $(-a, -b)$ and $x(-a, -b)$ for some real numbers a, b. We wish to prove that

$$(-a, -b) = -(a, b)$$
$$\text{and} \quad x(-a, -b) = x(a, b)$$

Case I, $(-a, -b)$ Let us start from the special case $(0, -a)$, for some real number a.

$$(0, -a) = -a$$
$$= -(0, a)$$

Thus

$$(0, -a) = -(0, a) \qquad (4.1)$$

Now, for some real numbers a, b,

$$(-a, -b) = (-a, 0) + (0, -b)$$
$$= -(0, -a) + (0, -b)$$

apply the result in (4.1)

$$= -[-(0, a)] - (0, b)$$
$$= (0, a) + (b, 0)$$
$$= (b, a)$$
$$\boxed{(-a, -b) = -(a, b)}$$

.

Case II, $x(-a, -b)$ Let us start from the special case $x(0, -a)$, for some real number a.

$$x(0, -a) = \frac{(-a)^2}{2}$$
$$= \frac{a^2}{2}$$
$$= x(0, a)$$

Thus

$$x(0, -a) = x(0, a) \qquad (4.2)$$

Now, for some real numbers a, b,

$$x(-a, -b) = x(-a, 0) + x(0, -b)$$
$$= -x(0, -a) + x(0, -b)$$

apply the result in (4.2)

$$= -x(0, a) + x(0, b)$$
$$= x(a, 0) + x(0, b)$$
$$\boxed{x(-a, -b) = x(a, b)}$$

4.2.2   Negative int-vectors when solving for a_n

The two types of negative int-vectors encountered when solving for a_n are $(-a, -b)$ and $x(-a, -b)$ for some real numbers a, b. We wish to prove that

$$(-a, -b) = -(a, b)$$
$$\text{and} \quad x(-a, -b) = x(a, b)$$

Case I, $(-a, -b)$ Let us start from the special case $(0, -a)$, for some real number a.

$$(0, -a) = \frac{1}{n\pi} \sin n(-a)$$

$$= -\frac{1}{n\pi} \sin na$$

$$= -(0, a)$$

Thus

$$(0, -a) = -(0, a) \tag{4.3}$$

Now, for some real numbers a, b,

$$(-a, -b) = (-a, 0) + (0, -b)$$

$$= -(0, -a) + (0, -b)$$

apply the result in (4.3)

$$= -[-(0, a)] - (0, b)$$

$$= (0, a) + (b, 0)$$

$$= (b, a)$$

$$\boxed{(-a, -b) = -(a, b)}$$

Case II, $x(-a, -b)$ Let us start from the special case $x(0, -a)$, for some real number a.

$$x(0, -a) = \frac{-a}{n\pi} \sin n(-a) + \frac{1}{n^2}[\cos n(-a) - 1]$$

$$= -(\frac{-a}{n\pi}) \sin na + \frac{1}{n^2}(\cos na - 1)$$

$$= \frac{a}{n\pi} \sin na + \frac{1}{n^2}(\cos na - 1)$$

$$= x(0, a)$$

Thus

$$x(0, -a) = x(0, a) \tag{4.4}$$

Now, for some real numbers a, b,

$$x(-a, -b) = x(-a, 0) + x(0, -b)$$

$$= -x(0, -a) + x(0, -b)$$

apply the result in (4.4)

$$= -x(0, a) + x(0, b)$$
$$= x(a, 0) + x(0, b)$$

$$\boxed{x(-a, -b) = x(a, b)}$$

4.2.3   Negative int-vectors when solving for b_n

The two types of negative int-vectors encountered when solving for b_n are $(-a, -b)$ and $x(-a, -b)$ for some real numbers a, b. It is easy to prove that

$$(-a, -b) = (a, b)$$
$$\text{and} \quad x(-a, -b) = -x(a, b)$$

The proof for these results is similar to the previous proof for a_n. Do this proof as an exercise.

4.3   Rules of Negative Int-Vectors

In this section, we shall collate all the results about negative int-vectors that we discussed in the last section. These results are now added to our list of Rules.

Rules when Solving for a_0 and a_n

For real numbers a, b,

Rule 5a $(-a, -b) = -(a, b)$

Rule 6a $x(-a, -b) = x(a, b)$

Rules when Solving for b_n

For real numbers $a, b,$

Rule 5b $(-a, -b) = (a, b)$

Rule 6b $x(-a, -b) = -x(a, b)$

Example 1

Assume you are solving for a_0 (or a_n) and simplify the following int-vectors.

1. $(-2, -3)$

2. $x(-1, -2)$

3. $(-6, 0)$

4. $x(0, -\pi)$

5. $(-\pi, -\frac{\pi}{2})$

6. $x(-\pi, -\frac{\pi}{2})$

7. $(-\frac{\pi}{2}, \frac{\pi}{2})$

8. $x(-\frac{\pi}{2}, \frac{\pi}{2})$

9. $(-3, 0) + (3, 0)$

Solution

1. $(-2, -3) = -(2, 3)$

2. $x(-1, -2) = x(1, 2)$

3. $(-6, 0) = -(6, 0)$

4. $x(0, -\pi) = x(0, \pi)$

5. $(-\pi, -\frac{\pi}{2}) = -(\pi, \frac{\pi}{2})$

6. $x(-\pi, -\frac{\pi}{2}) = x(\pi, \frac{\pi}{2})$

7.

$(-\frac{\pi}{2}, \frac{\pi}{2})$

$= (-\frac{\pi}{2}, 0) + (0, \frac{\pi}{2})$

apply Rule 5a

$= -(\frac{\pi}{2}, 0) + (0, \frac{\pi}{2})$

$= (0, \frac{\pi}{2}) + (0, \frac{\pi}{2})$

$= 2(0, \frac{\pi}{2})$

8.

$x(-\frac{\pi}{2}, \frac{\pi}{2})$

$= x(-\frac{\pi}{2}, 0) + x(0, \frac{\pi}{2})$

apply Rule 6a

$$= x(\frac{\pi}{2}, 0) + x(0, \frac{\pi}{2})$$

$$= -x(0, \frac{\pi}{2}) + x(0, \frac{\pi}{2})$$

$$= 0$$

9.

$$(-3, 0) + (3, 0)$$

$$= -(3, 0) + (3, 0)$$

$$= 0$$

Example 2

Assume you are solving for b_n and simplify the following int-vectors.

1. $(-2, -3)$

2. $x(-1, -2)$

3. $(-6, 0)$

4. $x(0, -\pi)$

5. $(-\pi, -\frac{\pi}{2})$

6. $x(-\pi, -\frac{\pi}{2})$

7. $(-\frac{\pi}{2}, \frac{\pi}{2})$

8. $x(-\frac{\pi}{2}, \frac{\pi}{2})$

9. $(-3, 0) + (3, 0)$

Solution

1. $(-2, -3) = (2, 3)$

2. $x(-1, -2) = -x(1, 2)$

3. $(-6, 0) = (6, 0)$

4. $x(0, -\pi) = -x(0, \pi)$

5. $(-\pi, -\frac{\pi}{2}) = (\pi, \frac{\pi}{2})$

6. $x(-\pi, -\frac{\pi}{2}) = -x(\pi, \frac{\pi}{2})$

7.

$$(-\frac{\pi}{2}, \frac{\pi}{2})$$

$$= (-\frac{\pi}{2}, 0) + (0, \frac{\pi}{2})$$

apply Rule 5b

$$= (\frac{\pi}{2}, 0) + (0, \frac{\pi}{2})$$

$$= -(0, \frac{\pi}{2}) + (0, \frac{\pi}{2})$$

$$= 0$$

8.

$$x(-\frac{\pi}{2}, \frac{\pi}{2})$$

$$= x(-\frac{\pi}{2}, 0) + x(0, \frac{\pi}{2})$$

apply Rule 6b

$$= -x(\frac{\pi}{2}, 0) + x(0, \frac{\pi}{2})$$

$$= x(0, \frac{\pi}{2}) + x(0, \frac{\pi}{2})$$

$$= 2x(0, \frac{\pi}{2})$$

9.

$$(-3, 0) + (3, 0)$$

$$= (3, 0) + (3, 0)$$

$$= 2(3, 0)$$

4.3.1 The Best Order to Solve for Fourier Coefficients

When we intend to get rid of negative values, by using Rule 5a to 6b in our working, it is much better to solve in this order:

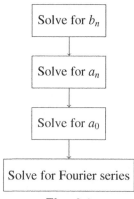

Fig. 4.1

Important Note

We shall make use of two reuse points.

reuse point 1 This point appears in *Solution for* b_n. It is the point immediately before we apply any Rules at all. It is the first place where all int-vectors are of the form $(0, a)$ or $x(0, a)$ for some real number a.

reuse point 2 This point appears in *Solution for* a_n. It is the point immediately before we apply Rule 3a and/or Rule 4a.

Example 1

Find the Fourier series for the function $f(x)$ which has a period of 2π.

$$f(x) = \begin{cases} -2\pi & -\pi < x < 0 \\ 3x & 0 < x < \pi \end{cases}$$

Solution for b_n

$b_n\pi =$ sum of cases

$\qquad = -2\pi(-\pi, 0) + 3x(0, \pi)$

$\qquad = 2\pi(0, -\pi) + 3x(0, \pi)$ $\hfill$ (reuse point 1)

apply Rule 5b

$\qquad = 2\pi(0, \pi) + 3x(0, \pi)$

apply Rules 3b and 4b

$$= \frac{2\pi}{n}(1 - \cos n\pi) - \frac{3\pi}{n}\cos n\pi + \frac{3}{n^2}\sin n\pi$$

$$= \frac{2\pi}{n}(1 - \cos n\pi) - \frac{3\pi}{n}\cos n\pi$$

$$= \frac{2\pi}{n} - \frac{2\pi}{n}\cos n\pi - \frac{3\pi}{n}\cos n\pi$$

$$b_n\pi = \frac{2\pi}{n} - \frac{5\pi}{n}\cos n\pi$$

$$b_n = \frac{2}{n} - \frac{5}{n}\cos n\pi$$

Therefore

$$b_n = \begin{cases} \dfrac{7}{n} & n \text{ is odd} \\ -\dfrac{3}{n} & n \text{ is even} \end{cases}$$

Solution for a_n   Continuing from reuse point 1,

$$a_n\pi = 2\pi(0, -\pi) + 3x(0, \pi)$$

apply Rule 5a

$$= -2\pi(0, \pi) + 3x(0, \pi) \qquad\qquad \text{(reuse point 2)}$$

apply Rules 3a and 4a

$$= -\frac{2\pi}{n}\sin n\pi + \frac{3\pi}{n}\sin n\pi + \frac{3}{n^2}(\cos n\pi - 1)$$

$$a_n\pi = \frac{3}{n^2}(\cos n\pi - 1)$$

$$a_n = \frac{3}{\pi n^2}(\cos n\pi - 1)$$

Therefore

$$a_n = \begin{cases} -\dfrac{6}{\pi n^2} & n \text{ is odd} \\[2mm] 0 & n \text{ is even} \end{cases}$$

Solution for a_0   Continuing from reuse point 2,

$$a_0\pi = -2\pi(0, \pi) + 3x(0, \pi)$$

apply Rules 3c and 4c

$$= -2\pi(\pi) + 3\frac{\pi^2}{2}$$

$$= -2\pi^2 + \frac{3\pi^2}{2}$$

$$a_0\pi = -\frac{\pi^2}{2}$$

$$a_0 = -\frac{\pi}{2}$$

Solution for Fourier series

$$f(x) = \frac{a_0}{2} + \sum_{n=1}^{\infty} a_n \cos nx + \sum_{n=1}^{\infty} b_n \sin nx$$

$$= -\frac{\pi}{4} - \frac{6}{\pi n^2}(\cos x + \cos 3x + \cdots) + 0(\cos 2x + \cos 4x + \cdots)$$

$$+ \frac{7}{n}(\sin x + \sin 3x + \cdots) - \frac{3}{n}(\sin 2x + \sin 4x + \cdots)$$

$$\boxed{\begin{array}{l} f(x) = -\frac{\pi}{4} - \frac{6}{\pi}(\cos x + \frac{1}{3^2}\cos 3x + \cdots) + 7(\sin x + \frac{1}{3}\sin 3x + \cdots) \\ \quad - 3(\frac{1}{2}\sin 2x + \frac{1}{4}\sin 4x + \cdots) \end{array}}$$

Example 2

Find the Fourier series for the function $f(x)$ which has a period of 2π.

$$f(x) = \begin{cases} \pi + x & -\pi < x < -\dfrac{\pi}{2} \\[2mm] \dfrac{\pi}{2} & -\dfrac{\pi}{2} < x < \dfrac{\pi}{2} \\[2mm] \pi - x & \dfrac{\pi}{2} < x < \pi \end{cases}$$

Solution for b_n

$$b_n \pi = \text{sum of cases}$$

$$= (\pi + x)(-\pi, -\frac{\pi}{2}) + \frac{\pi}{2}(-\frac{\pi}{2}, \frac{\pi}{2}) + (\pi - x)(\frac{\pi}{2}, \pi)$$

$$= \pi(-\pi, -\frac{\pi}{2}) + x(-\pi, -\frac{\pi}{2})$$

$$+ \frac{\pi}{2}(-\frac{\pi}{2}, \frac{\pi}{2})$$

$$+ \pi(\frac{\pi}{2}, \pi) - x(\frac{\pi}{2}, \pi) \qquad\qquad \text{(reuse point 1)}$$

apply Rules 5b and 6b

$$= \pi(\pi, \frac{\pi}{2}) - x(\pi, \frac{\pi}{2})$$

$$+ \frac{\pi}{2}(-\frac{\pi}{2}, 0) + \frac{\pi}{2}(0, \frac{\pi}{2})$$

$$+ \pi(\frac{\pi}{2}, \pi) - x(\frac{\pi}{2}, \pi)$$

$$= -\pi(\frac{\pi}{2}, \pi) + x(\frac{\pi}{2}, \pi)$$

$$+ \frac{\pi}{2}(\frac{\pi}{2}, 0) + \frac{\pi}{2}(0, \frac{\pi}{2})$$

$$+ \pi(\frac{\pi}{2}, \pi) - x(\frac{\pi}{2}, \pi)$$

rearrange the terms

$$= -\pi(\frac{\pi}{2}, \pi) + \pi(\frac{\pi}{2}, \pi)$$

$$- \frac{\pi}{2}(0, \frac{\pi}{2}) + \frac{\pi}{2}(0, \frac{\pi}{2})$$

$$+ x(\frac{\pi}{2}, \pi) - x(\frac{\pi}{2}, \pi)$$

$$b_n \pi = 0 + 0 + 0$$

$$\boxed{b_n = 0}$$

Solution for a_n   Continuing from reuse point 1,

$$a_n \pi = \pi(-\pi, -\frac{\pi}{2}) + x(-\pi, -\frac{\pi}{2})$$

$$+ \frac{\pi}{2}(-\frac{\pi}{2}, \frac{\pi}{2})$$

$$+ \pi(\frac{\pi}{2}, \pi) - x(\frac{\pi}{2}, \pi)$$

apply Rules 5a and 6a

$$= -\pi(\pi, \frac{\pi}{2}) + x(\pi, \frac{\pi}{2})$$

$$+ \frac{\pi}{2}(-\frac{\pi}{2}, 0) + \frac{\pi}{2}(0, \frac{\pi}{2})$$

$$+ \pi(\frac{\pi}{2}, \pi) - x(\frac{\pi}{2}, \pi)$$

$$= -\pi(\pi, 0) - \pi(0, \frac{\pi}{2}) + x(\pi, 0) + x(0, \frac{\pi}{2})$$

$$- \frac{\pi}{2}(\frac{\pi}{2}, 0) + \frac{\pi}{2}(0, \frac{\pi}{2})$$

$$+ \pi(\frac{\pi}{2}, 0) + \pi(0, \pi) - x(\frac{\pi}{2}, 0) - x(0, \pi)$$

$$= \pi(0, \pi) - \pi(0, \frac{\pi}{2}) - x(0, \pi) + x(0, \frac{\pi}{2})$$

$$+ \frac{\pi}{2}(0, \frac{\pi}{2}) + \frac{\pi}{2}(0, \frac{\pi}{2})$$

$$- \pi(0, \frac{\pi}{2}) + \pi(0, \pi) + x(0, \frac{\pi}{2}) - x(0, \pi)$$

$$= 2\pi(0, \pi) - 2\pi(0, \frac{\pi}{2}) + \pi(0, \frac{\pi}{2})$$

$$+ 2x(0, \frac{\pi}{2}) - 2x(0, \pi)$$

$$= 2\pi(0, \pi) - \pi(0, \frac{\pi}{2}) + 2x(0, \frac{\pi}{2}) - 2x(0, \pi) \qquad \text{(reuse point 2)}$$

apply Rules 3a and 4a

$$= \frac{2\pi}{n} \sin n\pi - \frac{\pi}{n} \sin \frac{n\pi}{2}$$

$$+ \frac{2\pi}{2n} \sin \frac{n\pi}{2} + \frac{2}{n^2}(\cos \frac{n\pi}{2} - 1)$$

$$- \frac{2\pi}{n} \sin n\pi - \frac{2}{n^2}(\cos n\pi - 1)$$

$$= 0 - \frac{\pi}{n} \sin \frac{n\pi}{2} + \frac{\pi}{n} \sin \frac{n\pi}{2} + \frac{2}{n^2}(\cos \frac{n\pi}{2} - 1)$$

$$- 0 - \frac{2}{n^2}(\cos n\pi - 1)$$

$$= 0 + \frac{2}{n^2}(\cos \frac{n\pi}{2} - 1) - \frac{2}{n^2}(\cos n\pi - 1)$$

$$a_n\pi = \frac{2}{n^2}(\cos \frac{n\pi}{2} - 1 - \cos n\pi + 1)$$

$$a_n = \frac{2}{\pi n^2}(\cos \frac{n\pi}{2} - \cos n\pi)$$

Therefore

$$a_n = \begin{cases} \dfrac{2}{\pi n^2} & n \text{ is odd} \\[2ex] -\dfrac{4}{\pi n^2} & n = 2, 6, 10, \ldots \\[2ex] 0 & n = 4, 8, 12, \ldots \end{cases}$$

Solution for a_0 Continuing from reuse point 2,

$$a_0\pi = 2\pi(0, \pi) - \pi(0, \frac{\pi}{2}) + 2x(0, \frac{\pi}{2}) - 2x(0, \pi)$$

apply Rules 3c and 4c

$$= 2\pi(\pi) - \pi\left(\frac{\pi}{2}\right) + 2\frac{(\pi/2)^2}{2} - 2\frac{(\pi)^2}{2}$$

$$= 2\pi^2 - \frac{\pi^2}{2} + \frac{\pi^2}{4} - \pi^2$$

$$a_0\pi = \pi^2\left(\frac{3}{4}\right)$$

$$\boxed{a_0 = \frac{3\pi}{4}}$$

Solution for Fourier Series

$$f(x) = \frac{a_0}{2} + \sum_{n=1}^{\infty} a_n \cos nx + \sum_{n=1}^{\infty} b_n \sin nx$$

$$= \frac{3\pi}{8} + a_n(\cos x + \cos 2x + \cos 3x + \cdots) + 0$$

$$= \frac{3\pi}{8} + a_n(\cos x + \cos 3x + \cdots) + a_n(\cos 2x + \cos 6x + \cdots)$$

$$+ a_n(\cos 4x + \cos 8x + \cdots)$$

$$= \frac{3\pi}{8} + \frac{2}{\pi n^2}(\cos x + \cos 3x + \cdots) - \frac{4}{\pi n^2}(\cos 2x + \cos 6x + \cdots)$$

$$+ 0(\cos 4x + \cos 8x + \cdots)$$

$$\boxed{f(x) = \frac{3\pi}{8} + \frac{2}{\pi}(\cos x + \frac{1}{3^2}\cos 3x + \cdots) - \frac{4}{\pi}(\frac{1}{2^2}\cos 2x + \frac{1}{6^2}\cos 6x + \cdots)}$$

Example 3

Find the Fourier series for the function $f(x)$ which has a period of 2π. The graph is as shown below.

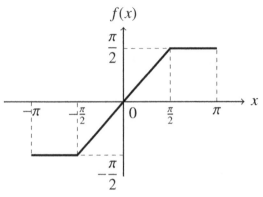

Solution for analytical definition   The analytical definition of $f(x)$ is

$$f(x) = \begin{cases} -\dfrac{\pi}{2} & -\pi < x < -\dfrac{\pi}{2} \\ x & -\dfrac{\pi}{2} < x < \dfrac{\pi}{2} \\ \dfrac{\pi}{2} & \dfrac{\pi}{2} < x < \pi \end{cases}$$

$$f(x + 2\pi) = f(x)$$

Solution for b_n

$b_n\pi = $ sum of cases

$$= -\frac{\pi}{2}(-\pi, -\frac{\pi}{2}) + x(-\frac{\pi}{2}, \frac{\pi}{2}) + \frac{\pi}{2}(\frac{\pi}{2}, \pi)$$

$$= -\frac{\pi}{2}(-\pi, 0) - \frac{\pi}{2}(0, -\frac{\pi}{2}) + x(-\frac{\pi}{2}, 0) + x(0, \frac{\pi}{2}) + \frac{\pi}{2}(\frac{\pi}{2}, 0) + \frac{\pi}{2}(0, \pi)$$

$$= \frac{\pi}{2}(0, -\pi) - \frac{\pi}{2}(0, -\frac{\pi}{2}) - x(0, -\frac{\pi}{2})$$

$$\quad + x(0, \frac{\pi}{2}) - \frac{\pi}{2}(0, \frac{\pi}{2}) + \frac{\pi}{2}(0, \pi) \qquad\qquad \text{(reuse point 1)}$$

apply Rules 5b and 6b

$$= \frac{\pi}{2}(0, \pi) - \frac{\pi}{2}(0, \frac{\pi}{2}) + x(0, \frac{\pi}{2})$$

$$\quad + x(0, \frac{\pi}{2}) - \frac{\pi}{2}(0, \frac{\pi}{2}) + \frac{\pi}{2}(0, \pi)$$

$$= \pi(0, \pi) - \pi(0, \frac{\pi}{2}) + 2x(0, \frac{\pi}{2})$$

apply Rules 3b and 4b

$$= \frac{\pi}{n}(1 - \cos n\pi) - \frac{\pi}{n}(1 - \cos \frac{n\pi}{2}) + 2\left(-\frac{\pi/2}{n} \cos \frac{n\pi}{2} + \frac{1}{n^2} \sin \frac{n\pi}{2} \right)$$

$$= \frac{\pi}{n} - \frac{\pi}{n} \cos n\pi - \frac{\pi}{n} + \frac{\pi}{n} \cos \frac{n\pi}{2} - \frac{\pi}{n} \cos \frac{n\pi}{2} + \frac{2}{n^2} \sin \frac{n\pi}{2}$$

$$b_n \pi = -\frac{\pi}{n} \cos n\pi + \frac{2}{n^2} \sin \frac{n\pi}{2}$$

$$b_n = -\frac{1}{n} \cos n\pi + \frac{2}{\pi n^2} \sin \frac{n\pi}{2}$$

Recall that

$$\sin \frac{n\pi}{2} = \begin{cases} 0 & n \text{ is even} \\ 1 & n = 1, 5, 9, \dots \\ -1 & n = 3, 7, 11, \dots \end{cases}$$

and

$$\cos n\pi = \begin{cases} -1 & n \text{ is odd} \\ 1 & n \text{ is even} \end{cases}$$

When n is even

$$b_n = -\frac{1}{n}(1) + \frac{2}{\pi n^2}(0)$$

$$= -\frac{1}{n}$$

When $n = 1, 5, 9, \dots$

$$b_n = -\frac{1}{n}(-1) + \frac{2}{\pi n^2}(1)$$

$$= \frac{1}{n} + \frac{2}{\pi n^2}$$

When $n = 3, 7, 11, \dots$

$$b_n = -\frac{1}{n}(-1) + \frac{2}{\pi n^2}(-1)$$

$$= \frac{1}{n} - \frac{2}{\pi n^2}$$

Therefore

$$b_n = \begin{cases} -\dfrac{1}{n} & n \text{ is even} \\[2mm] \dfrac{1}{n} + \dfrac{2}{\pi n^2} & n = 1, 5, 9, \ldots \\[2mm] \dfrac{1}{n} - \dfrac{2}{\pi n^2} & n = 3, 7, 11, \ldots \end{cases}$$

Solution for a_n   Continuing from reuse point 1,

$$a_n \pi = \frac{\pi}{2}(0, -\pi) - \frac{\pi}{2}(0, -\frac{\pi}{2}) - x(0, -\frac{\pi}{2})$$
$$+ x(0, \frac{\pi}{2}) - \frac{\pi}{2}(0, \frac{\pi}{2}) + \frac{\pi}{2}(0, \pi)$$

apply Rules 5a and 6a

$$= -\frac{\pi}{2}(0, \pi) + \frac{\pi}{2}(0, \frac{\pi}{2}) - x(0, \frac{\pi}{2})$$
$$+ x(0, \frac{\pi}{2}) - \frac{\pi}{2}(0, \frac{\pi}{2}) + \frac{\pi}{2}(0, \pi)$$

$$a_n \pi = 0$$

$$\boxed{a_n = 0}$$ (reuse point 2)

Solution for a_0   Continuing from reuse point 2,

$$\boxed{a_0 = 0}$$

Solution for Fourier series

$$f(x) = \frac{a_0}{2} + \sum_{n=1}^{\infty} a_n \cos nx + \sum_{n=1}^{\infty} b_n \sin nx$$

$$= 0 + 0(\cos x + \cos 2x + \cdots) + b_n(\sin x + \sin 2x + \cdots)$$

$$= b_n(\sin 2x + \sin 4x + \cdots) + b_n(\sin x + \sin 5x + \cdots)$$

$$+ b_n(\sin 3x + \sin 7x + \cdots)$$

$$= -\frac{1}{n}(\sin 2x + \sin 4x + \cdots) + (\frac{1}{n} + \frac{2}{\pi n^2})(\sin x + \sin 5x + \cdots)$$

$$+ (\frac{1}{n} - \frac{2}{\pi n^2})(\sin 3x + \sin 7x + \cdots)$$

$$= -\frac{1}{n}(\sin 2x + \sin 4x + \cdots) + \frac{1}{n}(\sin x + \sin 3x + \cdots)$$

$$+ \frac{2}{\pi n^2}(\sin x + \sin 5x + \cdots) - \frac{2}{\pi n^2}(\sin 3x + \sin 7x + \cdots)$$

$$= \frac{1}{n}(\sin x - \sin 2x + \cdots) + \frac{2}{\pi n^2}(\sin x - \sin 3x + \cdots)$$

$$f(x) = (\sin x - \frac{1}{2}\sin 2x + \cdots) + \frac{2}{\pi}(\sin x - \frac{1}{3^2}\sin 3x + \cdots)$$

Exercise 2

Find the Fourier series for the following functions. Assume they are periodic with period 2π.

1.

$$f(x) = |x| \qquad -\pi < x < \pi$$

2.

$$f(x) = \pi - |x| \qquad -\pi < x < \pi$$

3.

$$f(x) = \begin{cases} x + \pi & 0 < x < \pi \\ -x - \pi & -\pi < x < 0 \end{cases}$$

4.

$$f(x) = \begin{cases} -\pi & -\pi < x < 0 \\ x & 0 < x < \pi \end{cases}$$

5.

$$f(x) = \begin{cases} 0 & -\pi < x < 0 \\ 1 & 0 < x < \frac{\pi}{2} \\ 0 & \frac{\pi}{2} < x < \pi \end{cases}$$

6.

$$f(x) = \begin{cases} \pi + x & -\pi < x < -\frac{\pi}{2} \\ \frac{\pi}{2} & -\frac{\pi}{2} < x < \frac{\pi}{2} \\ \pi - x & \frac{\pi}{2} < x < \pi \end{cases}$$

Exercise 3

Find the Fourier series for the following functions. Assume they are periodic with period 2π.

1.

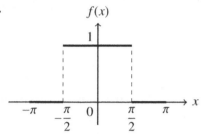

4.

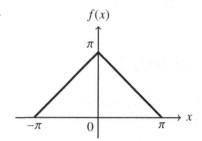

2.

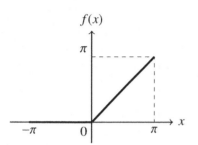

5.

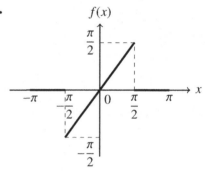

3.

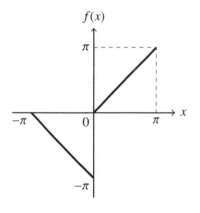

CHAPTER 5

Functions with Period T

In all previous problems where we expressed a function as a Fourier series, the period of the function was 2π. What about periodic functions with period other than 2π? How do we find the Fourier series of such functions? This will be the main discussion of this chapter.

In order to find the Fourier series of a function with period other than 2π, say T, we have to transform the underlying sine and cosine functions to have a period of T instead of 2π. Study the following example.

Example 1

The sine wave is shown below. It has a period of 2π as normal.

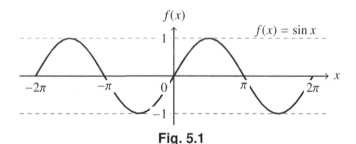

Fig. 5.1

We require a sine wave with period, say 5, instead of 2π. How do we get this? Simple. We only need to change the variable x to some other variable t such that

$$\frac{x}{2\pi} = \frac{t}{5}$$

Make x the subject of the formula.

$$x = \frac{2\pi}{5}t$$

Find sine of both sides

$$\sin x = \sin \frac{2\pi}{5}t$$

Thus $f(t) = \sin \dfrac{2\pi}{5}t$ is the required sine wave with period 5. The graph of $f(t)$ is shown below.

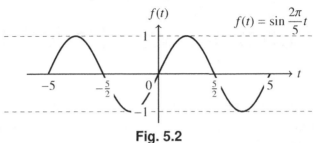

Fig. 5.2

Example 2

The periodic function shown in Fig. 5.3 has a period of 2π. Transform this function to a corresponding one with period of 2, given that:

$$f(x) = \sin x + \frac{1}{3}\sin 3x + \frac{1}{5}\sin 5x \qquad (5.1)$$

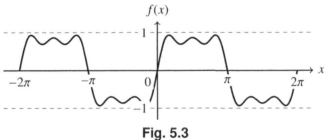

Fig. 5.3

Solution   The required period is 2. Thus

$$\frac{x}{2\pi} = \frac{t}{2}$$

$$x = \pi t$$

put $x = \pi t$ in (5.1)

$$\therefore f(t) = \sin \pi t + \frac{1}{3}\sin 3\pi t + \frac{1}{5}\sin 5\pi t$$

If we graph $f(t)$ we get:

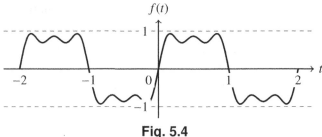

Fig. 5.4

5.1 Generalization

In general, to transform a function $f(x)$ with period 2π to a function with period T, we simply change the variable x to another variable t that satisfies the equation:

$$\frac{x}{2\pi} = \frac{t}{T}$$

Thus we only need to substitute $x = \dfrac{2\pi}{T}$ in $f(x)$ to transform it to a function with period T.

Since the term $\dfrac{2\pi}{T}$ occurs frequently, we shall denote it with the symbol ω. Thus $\omega = \dfrac{2\pi}{T}$ and $x = \omega t$.

5.2 Fourier Series

In Chapter 3, it was stated that the Fourier series of a function $f(x)$ with period 2π is:

$$f(x) = \frac{a_0}{2} + \sum_{n=1}^{\infty} a_n \cos nx + \sum_{n=1}^{\infty} b_n \sin nx$$

Now we wish to generalize this and transform $f(x)$ to a periodic function of arbitrary period T. As we discussed in the previous section, this requires us to substitute $x = \omega t$ in $f(x)$. Therefore

$$f(t) = \frac{a_0}{2} + \sum_{n=1}^{\infty} a_n \cos n\omega t + \sum_{n=1}^{\infty} b_n \sin n\omega t$$

where $\omega = \dfrac{2\pi}{T}$.

This is the Fourier series expansion of a periodic function of period T. However, since we prefer x to be the variable in most mathematics, in this text the standard form of the Fourier series of a function, with period T, is

$$f(x) = \frac{a_0}{2} + \sum_{n=1}^{\infty} a_n \cos n\omega x + \sum_{n=1}^{\infty} b_n \sin n\omega x$$

where $\omega = \dfrac{2\pi}{T}$.

5.2.1   The New Rules

When solving for Fourier series of functions with period other than 2π we shall use some slightly different rules. All rules of int-vectors remain the same except Rules 3a, 4a, 3b and 4b; these rules are changed to the more general rules shown below. Of course, these rules can also be used to solve functions with period 2π, as they are a special case.

Rules when solving for a_n

New Rule 3a $(0, a) = \dfrac{1}{n\omega} \sin n\omega a$

New Rule 4a $x(0, a) = \dfrac{a}{n\omega} \sin n\omega a + \dfrac{1}{n^2\omega^2}(\cos n\omega a - 1)$

Rules when solving for b_n

New Rule 3b $(0, a) = \dfrac{1}{n\omega}(1 - \cos n\omega a)$

New Rule 4b $x(0, a) = -\dfrac{a}{n\omega} \cos n\omega a + \dfrac{1}{n^2\omega^2} \sin n\omega a$

Remember to make a note of these results. They will be used frequently when solving for Fourier series problems.

Important Note   Henceforth, Rules 3a, 4a, 3b and 4b shall refer to the newer versions of the respective Rules.

5.2.2   Routine for Solving Problems

This is the new routine for solving Fourier series problems. Take note of the few changes.

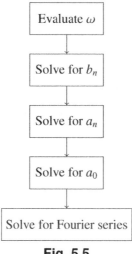

Fig. 5.5

1. **Evaluate** ω: Evaluate the value of ω from the given period T. The formula is:

$$\omega = \frac{2\pi}{T}$$

2. **Solve for** b_n: The formula for b_n is given as:

$$b_n \frac{\pi}{\omega} = \text{sum of cases}$$

Apply only the General rules to the int-vectors until all int-vectors are of the form $(0, a)$ or $x(0, a)$ for some real number a. This point in the working is labeled as *reuse point 1*. Then apply Rule 3b and/or Rule 4b to get the value of b_n.

3. **Solve for** a_n: Solve for a_n by continuing from *reuse point 1*. Use Rules 5a and 6a to eliminate negative numbers, if present, from the int-vectors. The line just before you apply Rule 3a and/or Rule 4a is labeled as *reuse point 2*. Apply the rules to get the value of a_n.

4. Solve for a_0: Solve for a_0 by continuing from *reuse point 2*. Apply Rule 3c and/or Rule 4c to get the value of a_0.

However, if you are given the graph of the function, you can also solve for a_0 by determining the area (A) of one period of the function. The Fourier coefficient a_0 is given by the formula:

$$a_0 \frac{\pi}{\omega} = A$$

5. Solve for Fourier series: The Fourier series for a function $f(x)$ is given as:

$$f(x) = \frac{a_0}{2} + \sum_{n=1}^{\infty} a_n \cos n\omega x + \sum_{n=1}^{\infty} b_n \sin n\omega x$$

where $\omega = \dfrac{2\pi}{T}$.

Simply substitute a_0, a_n and b_n with their respective values according to what has been solved.

Example 1

Find the Fourier series for the function defined by

$$f(x) = \begin{cases} 1 & -1 < x < 0 \\ x & 0 < x < 1 \end{cases}$$

$$f(x + 2) = f(x)$$

Solution for ω   The period T is given as 2.

$$\omega = \frac{2\pi}{T}$$

$$= \frac{2\pi}{2} = \pi$$

Solution for b_n

$$b_n \frac{\pi}{\omega} = \text{sum of cases}$$

$$= 1(-1, 0) + x(0, 1)$$

put $\omega = \pi$

$$b_n = 1(-1, 0) + x(0, 1) \qquad \text{(reuse point 1)}$$

apply Rule 5b

$$= (1, 0) + x(0, 1)$$
$$= -(0, 1) + x(0, 1)$$

apply Rules 3b and 4b

$$= -\frac{1}{n\omega}[1 - \cos n\omega(1)] + [-\frac{1}{n\omega} \cos n\omega(1) + \frac{1}{n^2\omega^2} \sin n\omega(1)]$$

put $\omega = \pi$

$$= -\frac{1}{n\pi}(1 - \cos n\pi) + [-\frac{1}{n\pi} \cos n\pi + \frac{1}{n^2\pi^2} \sin n\pi]$$

$$= -\frac{1}{n\pi}(1 - \cos n\pi) - \frac{1}{n\pi} \cos n\pi + 0$$

$$= -\frac{1}{n\pi} + \frac{1}{n\pi} \cos n\pi - \frac{1}{n\pi} \cos n\pi$$

$$\boxed{b_n = -\frac{1}{n\pi}}$$

Solution for a_n   Continuing from reuse point 1,

$$a_n = 1(-1, 0) + x(0, 1)$$

apply Rule 5a

$$= -(1, 0) + x(0, 1)$$
$$= (0, 1) + x(0, 1) \qquad \text{(reuse point 2)}$$

apply Rules 3a and 4a

$$= \frac{1}{n\omega} \sin n\omega(1) + [\frac{1}{n\omega} \sin n\omega(1) + \frac{1}{n^2\omega^2}(\cos n\omega(1) - 1)]$$

put $\omega = \pi$

$$= \frac{1}{n\pi} \sin n\pi + [\frac{1}{n\pi} \sin n\pi + \frac{1}{n^2\pi^2}(\cos n\pi - 1)]$$

$$= 0 + 0 + \frac{1}{n^2\pi^2}(\cos n\pi - 1)$$

$$a_n = \frac{1}{n^2\pi^2}(\cos n\pi - 1)$$

Therefore

$$a_n = \begin{cases} -\dfrac{2}{n^2\pi^2} & n \text{ is odd} \\ \\ 0 & n \text{ is even} \end{cases}$$

Solution for a_0   Continuing from reuse point 2,

$$a_0 = (0, 1) + x(0, 1)$$

apply Rules 3c and 4c

$$= 1 + \frac{1^2}{2}$$

$$\boxed{a_0 = \frac{3}{2}}$$

Solution for Fourier series

$$f(x) = \frac{a_0}{2} + \sum_{n=1}^{\infty} a_n \cos n\omega x + \sum_{n=1}^{\infty} b_n \sin n\omega x$$

$$= \frac{3}{4} + a_n(\cos \pi x + \cos 2\pi x + \cos 3\pi x + \cdots)$$

$$+ b_n(\sin \pi x + \sin 2\pi x + \sin 3\pi x + \cdots)$$

$$= \frac{3}{4} + a_n(\cos \pi x + \cos 3\pi x + \cdots) + a_n(\cos 2\pi x + \cos 4\pi x + \cdots)$$

$$+ b_n(\sin \pi x + \sin 2\pi x + \sin 3\pi x + \cdots)$$

$$= \frac{3}{4} - \frac{2}{n^2\pi^2}(\cos \pi x + \cos 3\pi x + \cdots) + 0(\cos 2\pi x + \cos 4\pi x + \cdots)$$

$$- \frac{1}{n\pi}(\sin \pi x + \sin 2\pi x + \sin 3\pi x + \cdots)$$

$$f(x) = \frac{3}{4} - \frac{2}{\pi^2}\left(\cos \pi x + \frac{1}{3^2}\cos 3\pi x + \cdots\right) - \frac{1}{\pi}\left(\sin \pi x + \frac{1}{2}\sin 2\pi x + \cdots\right)$$

Example 2

Find the Fourier series for the function defined by

$$f(x) = \begin{cases} 0 & -2 < x < -1 \\ -2 & -1 < x < 0 \\ 1 & 0 < x < 1 \\ 0 & 1 < x < 2 \end{cases}$$

$$f(x + 4) = f(x)$$

Solution for ω   The period T is given as 4.
   Now,

$$\omega = \frac{2\pi}{T}$$

$$= \frac{2\pi}{4} = \frac{\pi}{2}$$

Solution for b_n

$$b_n\frac{\pi}{\omega} = \text{sum of cases}$$

$$\text{put } \omega = \frac{\pi}{2}$$

$$b_n\frac{\pi}{\pi/2} = \text{sum of cases}$$

$$2b_n = \text{sum of cases}$$

$$= 0(-2, -1) - 2(-1, 0) + 1(0, 1) + 0(1, 2)$$

$$= -2(-1, 0) + (0, 1) \qquad \text{(reuse point 1)}$$

apply Rule 5b

$$= -2(1, 0) + (0, 1)$$
$$= 2(0, 1) + (0, 1)$$
$$= 3(0, 1)$$

apply Rule 3b

$$= 3\frac{1}{n\omega}[1 - \cos n\omega(1)]$$

put $\omega = \dfrac{\pi}{2}$

$$2b_n = 3\frac{2}{n\pi}(1 - \cos\frac{n\pi}{2})$$

$$b_n = \frac{3}{n\pi}(1 - \cos\frac{n\pi}{2})$$

Recall that

$$\cos\frac{n\pi}{2} = \begin{cases} 0 & n \text{ is odd} \\ -1 & n = 2, 6, 10, \ldots \\ 1 & n = 4, 8, 12, \ldots \end{cases}$$

Therefore

$$b_n = \begin{cases} \dfrac{3}{n\pi} & n \text{ is odd} \\ \dfrac{6}{n\pi} & n = 2, 6, 10, \ldots \\ 0 & n = 4, 8, 12, \ldots \end{cases}$$

Solution for a_n   Continuing from reuse point 1,

$$2a_n = -2(-1, 0) + (0, 1)$$

apply Rule 5a

$$= 2(1, 0) + (0, 1)$$
$$= -2(0, 1) + (0, 1)$$
$$= -(0, 1) \qquad\qquad \text{(reuse point 2)}$$

apply Rule 3a

$$= -\frac{1}{n\omega}\sin n\omega(1)$$

put $\omega = \dfrac{\pi}{2}$

$$= -\frac{1}{n(\pi/2)} \sin \frac{n\pi}{2}$$

$$2a_n = -\frac{2}{n\pi} \sin \frac{n\pi}{2}$$

$$a_n = -\frac{1}{n\pi} \sin \frac{n\pi}{2}$$

Recall that,

$$\sin \frac{n\pi}{2} = \begin{cases} 0 & n \text{ is even} \\ 1 & n = 1, 5, 9, \ldots \\ -1 & n = 3, 7, 11, \ldots \end{cases}$$

Therefore,

$$\boxed{a_n = \begin{cases} 0 & n \text{ is even} \\ -\dfrac{1}{n\pi} & n = 1, 5, 9, \ldots \\ \dfrac{1}{n\pi} & n = 3, 7, 11, \ldots \end{cases}}$$

Solution for a_0   Continuing from reuse point 2,

$$2a_0 = -(0, 1)$$

apply Rule 3c

$$2a_0 = -1$$

$$\boxed{a_0 = -\frac{1}{2}}$$

Solution for Fourier series

$$f(x) = \frac{a_0}{2} + \sum_{n=1}^{\infty} a_n \cos n\omega x + \sum_{n=1}^{\infty} b_n \sin n\omega x$$

$$= -\frac{1}{4} + a_n \left(\cos \frac{\pi x}{2} + \cos 2\frac{\pi x}{2} + \cos 3\frac{\pi x}{2} + \cdots \right)$$

$$+ b_n \left(\sin \frac{\pi x}{2} + \sin 2\frac{\pi x}{2} + \sin 3\frac{\pi x}{2} + \cdots \right)$$

$$= -\frac{1}{4} + 0 \left(\cos 2\frac{\pi x}{2} + \cos 4\frac{\pi x}{2} + \cdots \right) - \frac{1}{n\pi} \left(\cos \frac{\pi x}{2} + \cos 5\frac{\pi x}{2} + \cdots \right)$$

$$+ \frac{1}{n\pi}(\cos 3\frac{\pi x}{2} + \cos 7\frac{\pi x}{2} + \cdots)$$

$$+ \frac{3}{n\pi}(\sin \frac{\pi x}{2} + \sin 3\frac{\pi x}{2} + \cdots) + \frac{6}{n\pi}(\sin 2\frac{\pi x}{2} + \sin 6\frac{\pi x}{2} + \cdots)$$

$$+ 0(\sin 4\frac{\pi x}{2} + \sin 8\frac{\pi x}{2} + \cdots)$$

$$= -\frac{1}{4} - \frac{1}{\pi}(\cos \frac{\pi x}{2} + \frac{1}{5}\cos 5\frac{\pi x}{2} + \cdots) + \frac{1}{\pi}(\frac{1}{3}\cos 3\frac{\pi x}{2} + \frac{1}{7}\cos 7\frac{\pi x}{2} + \cdots)$$

$$+ \frac{3}{\pi}(\sin \frac{\pi x}{2} + \frac{1}{3}\sin 3\frac{\pi x}{2} + \cdots) + \frac{6}{\pi}(\frac{1}{2}\sin 2\frac{\pi x}{2} + \frac{1}{6}\sin 6\frac{\pi x}{2} + \cdots)$$

$$\boxed{\begin{aligned} f(x) &= -\frac{1}{4} - \frac{1}{\pi}(\cos \frac{\pi x}{2} - \frac{1}{3}\cos 3\frac{\pi x}{2} + \cdots) \\ &\quad + \frac{3}{\pi}(\sin \frac{\pi x}{2} + \frac{1}{3}\sin 3\frac{\pi x}{2} + \cdots) + \frac{6}{\pi}(\frac{1}{2}\sin 2\frac{\pi x}{2} + \frac{1}{6}\sin 6\frac{\pi x}{2} + \cdots) \end{aligned}}$$

Exercise

Find the Fourier series for the following functions.

1.

$$f(x) = \begin{cases} 0 & -1 < x < 0 \\ x & 0 < x < 1 \end{cases}$$

$$f(x + 2) = f(x)$$

4.

$$f(x) = \begin{cases} x + 5 & -2 < x < 0 \\ -x + 5 & 0 < x < 2 \end{cases}$$

$$f(x + 4) = f(x)$$

2.

$$f(x) = \begin{cases} 1 & -5 < x < 0 \\ 1 + x & 0 < x < 5 \end{cases}$$

$$f(x + 10) = f(x)$$

5.

$$f(x) = \begin{cases} \pi & -1 < x < 0 \\ -\pi & 0 < x < 1 \end{cases}$$

$$f(x + 2) = f(x)$$

3.

$$f(x) = \begin{cases} 2 + x & -2 < x < 0 \\ 2 & 0 < x < 2 \end{cases}$$

$$f(x + 4) = f(x)$$

6.

$$f(x) = 1 - x \quad 0 < x < 2$$

$$f(x + 2) = f(x)$$

7.

$$f(x) = \begin{cases} 1 & -2 < x < -1 \\ 0 & -1 < x < 1 \\ 1 & 1 < x < 2 \end{cases}$$

$$f(x+4) = f(x)$$

10.

$$f(x) = \begin{cases} 1 & 0 < x < \dfrac{1}{2} \\ 0 & \dfrac{1}{2} < x < 1 \end{cases}$$

$$f(x+1) = f(x)$$

8.

$$f(x) = \begin{cases} x+1 & -1 < x < 0 \\ x-1 & 0 < x < 1 \end{cases}$$

$$f(x+2) = f(x)$$

11.

$$f(x) = \begin{cases} 0 & 0 < x < \dfrac{1}{2} \\ 1 & \dfrac{1}{2} < x < 1 \end{cases}$$

$$f(x+1) = f(x)$$

9.

$$f(x) = \begin{cases} 1 & -2 < x < -1 \\ -x & -1 < x < 0 \\ x & 0 < x < 1 \\ 1 & 1 < x < 2 \end{cases}$$

$$f(x+4) = f(x)$$

CHAPTER 6

Odd and Even Functions

Functions can be classified as odd, even or neither. We can determine this property of a function based on the shape of its graph; particularly its symmetry. A basic knowledge of this property of functions can save the labor of finding the Fourier coefficients in a problem.

The knowledge of odd and even functions will be required in Chapter 7 where we expand a non-periodic function as a Fourier series made up entirely of sines or cosines.

6.1   Even Functions

An even function is one whose graph is *symmetrical about the y-axis*. This means if you flip the graph from right to left, the shape of the graph remains the same. The technical term for 'flip from right to left' is 'reflect about the y-axis'. An example of such a graph is shown in Fig. 6.1. If we flip this graph *from right to left* we get the graph in Fig. 6.2.

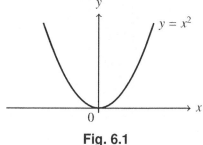

Fig. 6.1

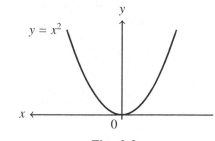

Fig. 6.2

The half of the graph at the left of the y-axis and the other half at the right of the y-axis are mirror images of each other.

Another example of an even function is the function $y = \cos x$. See Fig. 6.3.

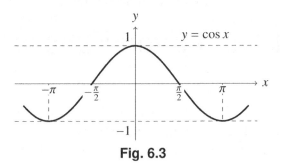

Fig. 6.3

Note   An even function does not have to be periodic. The function $\cos x$ is periodic, whereas the function x^2 is not periodic.

6.1.1   How to Determine an Even Graph by Comparing with Real Life Examples

There is symmetry everywhere in nature. This means we can find objects similar to even graphs in real life. Common examples are the human face, a heart-shape, etc.

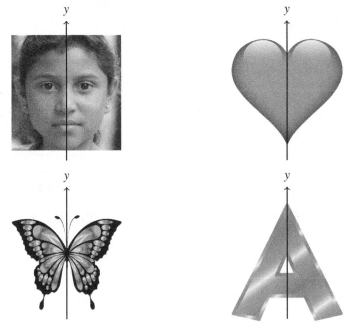

Fig. 6.4

Compare these images with the even graph in Fig. 6.5. Notice how they have the same symmetry.

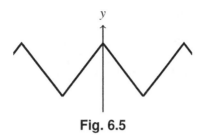

Fig. 6.5

Important Note When determining whether or not a graph is even, you should consider whether the y-axis divides the shape into two halves that are mirror images of each other. If it does, then the graph is even, else it is not. The x-axis should be ignored. This is why the previous even graph had no x-axis.

6.1.2 Theorem of Even Functions

Due to the symmetry of the graph of an even function $f(x)$ about the y-axis, the following equation is always true for even functions.

$$f(-x) = f(x)$$

That is, the function value for a particular negative value of x is same as that for the corresponding positive value of x.

Example 1

The function $y = x^2$ is an even function, therefore

$$(-2)^2 = 4 = (2)^2$$
$$(-3)^2 = 9 = (3)^2 \quad \text{etc.}$$

Example 2

The function $y = \cos x$ is an even function, therefore

$$\cos(-\pi) = -1 = \cos(\pi)$$
$$\cos(-\frac{\pi}{2}) = 0 = \cos(\frac{\pi}{2}) \quad \text{etc.}$$

6.2   Odd Functions

This is a function whose graph is *symmetrical about the origin*. What does this mean? It simply means if you flip the graph from right to left and then from top to bottom, the shape of the graph remains unchanged. The technical term for 'flip from right to left' is 'reflect about the y-axis' and the technical term for 'flip from top to bottom' is 'reflect about the x-axis'

An example of an odd function is $y = \sin x$. The graph is shown below.

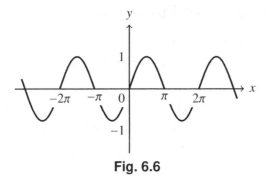

Fig. 6.6

If we flip this graph *from right to left* we get:

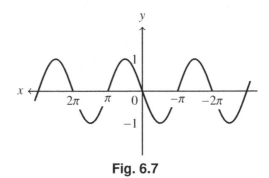

Fig. 6.7

Notice how the x-axis is now pointing left.

Now we shall flip this new graph *from top to bottom.*

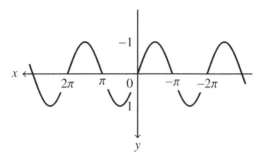

Fig. 6.8

Notice how the *y*-axis is now pointing downwards. We ended up with the shape we started with. This proves that $f(x) = \sin x$ is an odd function.

Another example of an odd function is $y = x^3$. The graph is shown below.

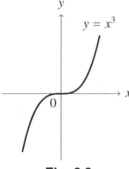

Fig. 6.9

The figure below shows the different flips of the graph of $y = x^3$. It proves that $y = x^3$ is an odd function.

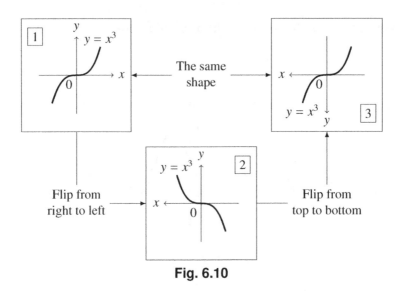

Fig. 6.10

Note   An odd function does not have to be periodic. The function $\sin x$ is periodic, whereas the function x^3 is not periodic.

6.2.1   How to Determine whether a Graph is Odd by Rotation

You can easily tell whether a graph is odd by rotating the graph about the origin.

Now look at the graph shown below.

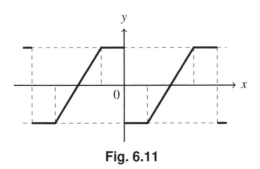

Fig. 6.11

Turn this book upside-down and look at the graph again. What do you notice? The graph did not change. The only difference is that after rotation, the x-axis is now pointing left while the y-axis is now pointing downwards. This actually proves that the graph is odd.

But what just happened? When you turned this book upside-down, you actually rotated the graph about the origin by exactly 180°. Therefore we can state that *for every odd function, rotating the graph about the origin by exactly 180° does not change the shape of the graph.*

6.2.2   Theorem of Odd Functions

Due to the symmetry of the graph of an odd function $f(x)$ about the origin, the following theorem is always true for odd functions.

$$f(-x) = -f(x)$$

That is, the function value for a particular negative value of x is numerically equal to that for the corresponding positive value of x but opposite in sign.

Example 1

The function $y = x^3$ is an odd function, therefore

$$(-2)^3 = -8 = -(2)^3$$
$$(-3)^3 = -27 = -(3)^3 \quad \text{etc.}$$

Example 2

The function $y = \sin x$ is an odd function, therefore

$$\sin(-\frac{\pi}{2}) = -1 = -\sin(\frac{\pi}{2})$$
$$\sin(-\frac{3\pi}{2}) = 1 = -\sin(\frac{3\pi}{2}) \quad \text{etc.}$$

6.3   Neither Even nor Odd Functions

A function may be neither even nor odd. This means the graph of the function does not possess symmetry about the y-axis and does not possess symmetry about the origin. They are quite common in Fourier series. An example of such function is shown below.

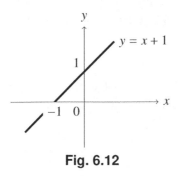

Fig. 6.12

Let us prove it is neither even nor odd in two stages. First we shall prove it is not even and second we shall prove it is not odd.

To check whether it is even we can flip it from right to left and compare this new graph with the initial one. This is shown below.

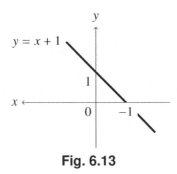

Fig. 6.13

As you can see, this new graph has a totally different shape. Thus the function is not even.

Now let us check whether the function is odd. To do this we can continue from the last graph as it is part of the process of determining whether a graph is odd. If we flip the last graph from top to bottom we get:

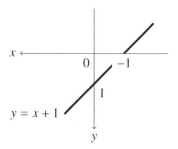

Fig. 6.14

This graph is not the same as the graph of the function. Thus the function is not odd. Therefore the function is *neither odd nor even*.

Notice that since we started by checking for the even property, we could reuse the resulting graph in checking whether the graph is odd. Thus the whole process was seamless. This method is visualized in a flowchart in the next section.

Both Even and Odd   There is only one function that is both even and odd. This is the function $f(x) = 0$. It possesses symmetry about the y-axis and about the origin.

Example

State whether each of functions below is odd, even or neither.

1.

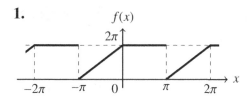

2.

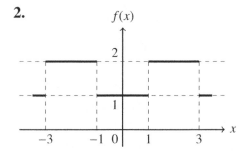

3.

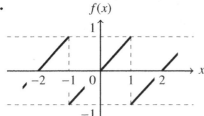

4.

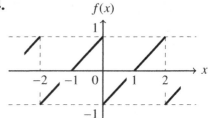

5.

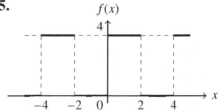

Solution

1. Neither odd nor even

2. Even

3. Odd

4. Odd

5. Neither odd nor even

6.4   Flowchart to Determine whether a Function is Even, Odd or Neither

This flowchart is not for a quick solution to determine the property of a function. It is for demonstrating in an efficient and error-free manner that a function has a particular property.

The key to the flowchart is

- *A* means the graph of $f(x)$,

- *B* means the result of flipping *A* from right to left,

- *C* means the result of flipping *B* from top to bottom.

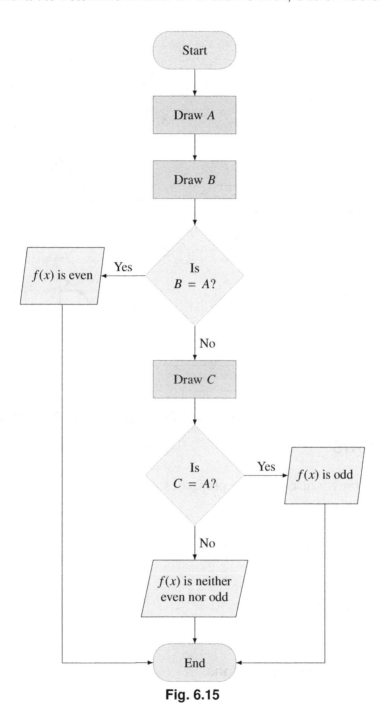

Fig. 6.15

Exercise 1

State whether each of the following functions is odd, even or neither.

1.

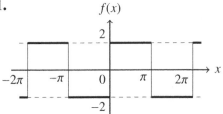

3.

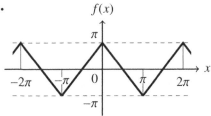

2.

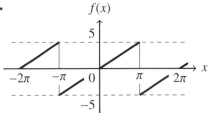

4.

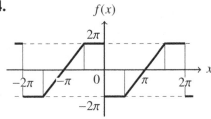

6.5   Products and Sums of Odd and Even Functions

In this section we discuss some properties of products and sums of odd and even functions.

Products of Odd and Even Functions

$(\text{even}) \times (\text{even}) = (\text{even})$

$(\text{odd}) \times (\text{odd}) = (\text{even})$

$(\text{even}) \times (\text{odd}) = (\text{odd})$

Sum of Odd and Even Functions

$(\text{even}) + (\text{even}) = (\text{even})$

$(\text{odd}) + (\text{odd}) = (\text{odd})$

$(\text{even}) + (\text{odd}) = (\text{neither even nor odd})$

There is an exception to the last rule. It does not hold when one of the functions is $f(x) = 0$.

6.6 How to Determine whether a Function is Odd, Even or Neither from its Fourier Series

We can determine whether a function is odd, even or neither from its Fourier series alone. To do this we shall simply apply the rules for product and sum of odd and even functions.

Now, $\sin x$, $\sin 2x$, $\sin 3x$, etc., are all odd functions. Thus we can say $\sin kx$ is odd for any real number k. Similarly, $\cos x$, $\cos 2x$, $\cos 3x$, etc., are all even functions. Thus we can say $\cos kx$ is even for any real number k.

Let E denote even functions and D odd functions, such that $f(x) = E$ simply means $f(x)$ is even and $f(x) = D$ simply means $f(x)$ is odd. Also, $f(x) = E + D$ implies $f(x)$ is neither even nor odd, provided $E \neq 0$ and $D \neq 0$. (We need $E \neq 0$ and $D \neq 0$ because $f(x) = 0$ is both even and odd.) Therefore we can write

$$\sin kx = D$$
$$\cos kx = E$$
$$\text{constant} = E$$

Therefore

$$K \sin kx = D$$
$$K \cos kx = E$$

for real numbers K, k

In the above equation, *constant* refers to the function $f(x) = \text{constant}$, which is an even function.

By using this notation, we can decompose any function into component even or odd functions and apply the rules of product and sums of functions to determine whether it is odd, even or neither.

Example 1

Determine whether the function below is odd, even or neither.

$$f(x) = \frac{4}{\pi}\left(\sin x + \frac{1}{3}\sin 3x + \cdots \right)$$

Solution   Now $\frac{4}{\pi}$ is a constant thus $\frac{4}{\pi} = E$. Also, $K \sin kx = D$. Thus

$$\begin{aligned} f(x) &= \frac{4}{\pi}\left(\sin x + \frac{1}{3}\sin 3x + \cdots \right) \\ &= E\left(D + D + \cdots\right) \\ &= E \times D \\ &= D \end{aligned}$$

Therefore $f(x)$ is an odd function.

Example 2

Determine whether the function below is odd, even or neither.

$$f(x) = \frac{\pi}{4} - \frac{2}{\pi}\left(\cos x + \frac{1}{3^2}\cos 3x + \cdots \right) - \left(\sin x - \frac{1}{2}\sin 2x + \cdots \right)$$

Solution

$$\begin{aligned} f(x) &= \frac{\pi}{4} + \left(-\frac{2}{\pi}\right)\left(\cos x + \frac{1}{3^2}\cos 3x + \cdots \right) + (-1)\left(\sin x - \frac{1}{2}\sin 2x + \cdots \right) \\ &= E + E\left(E + E + \cdots\right) + E\left(D + D + \cdots\right) \\ &= E + E \times E + E \times D \\ &= E + E + D \\ &= E + D \qquad \text{where } E \neq 0 \text{ and } D \neq 0 \end{aligned}$$

Therefore $f(x)$ is neither odd nor even.

Exercise 2

Determine whether the following functions are even, odd or neither.

1. $f(x) = \dfrac{\pi}{2} + 2\left(\sin x + \dfrac{1}{3} \sin 3x + \cdots \right)$

2. $f(x) = 2\left(\sin x + \dfrac{1}{2} \sin 2x + \cdots \right)$

3. $f(x) = \dfrac{\pi}{2} - \dfrac{4}{\pi}\left(\cos x + \dfrac{1}{3^2} \cos 3x + \cdots \right)$

4. $f(x) = \dfrac{3\pi}{2} - \dfrac{2}{\pi}\left(\cos x + \dfrac{1}{3^2} \cos 3x + \cdots \right) - \left(\sin x - \dfrac{1}{2} \sin 2x + \cdots \right)$

5. $f(x) = 2\left(\cos x - \dfrac{1}{3} \cos 3x + \cdots \right) + \dfrac{4}{\pi}\left(\dfrac{1}{2} \sin 2x + \dfrac{1}{6} \sin 6x + \cdots \right)$

6.7 Fourier Series of Odd and Even Functions

We can take advantage of the odd or even property of a periodic function to find its Fourier series with less work.

Even Function   The Fourier series of an even function has to be the sum of other even functions. Since $f(x) = \sin x$, $\sin 2x$, $\sin 3x$ etc., are all odd functions they cannot be part of the Fourier series. Therefore, the Fourier series of an even function contains only cosine terms and possibly a constant $\dfrac{a_0}{2}$. Notice that $f(x) = $ constant is also an even function.

Thus for an even function,

$$\boxed{b_n = 0}$$

Odd Function   The Fourier series of an odd function has to be the sum of other odd functions. Since $f(x) =$ constant, $\cos x$, $\cos 2x$ etc., are all even functions they cannot be part of the Fourier series. Therefore, the Fourier series of an odd function contains only sine terms.

Thus for an odd function,

$$\boxed{a_n = 0} \quad \text{and} \quad \boxed{a_0 = 0}$$

Example 1

Find the Fourier series for the function $f(x)$ given that the graph is as shown below. Assume period is 2π.

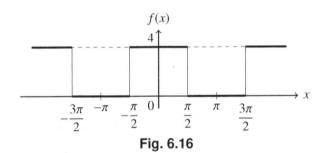

Fig. 6.16

Determine the property of $f(x)$ The graph of the function is symmetrical about the y-axis. Thus it is even. Therefore

$$\boxed{b_n = 0}$$

Solution for analytical definition   We can define the function analytically as:

$$f(x) = \begin{cases} 0 & -\pi < x < -\dfrac{\pi}{2} \\[2mm] 4 & -\dfrac{\pi}{2} < x < \dfrac{\pi}{2} \\[2mm] 0 & \dfrac{\pi}{2} < x < \pi \end{cases}$$

$$f(x + 2\pi) = f(x)$$

Solution for a_n

$$a_n\pi = \text{sum of cases}$$

$$= 0(-\pi, -\frac{\pi}{2}) + 4(-\frac{\pi}{2}, \frac{\pi}{2}) + 0(\frac{\pi}{2}, \pi)$$

$$= 4(-\frac{\pi}{2}, \frac{\pi}{2})$$

$$= 4(-\frac{\pi}{2}, 0) + 4(0, \frac{\pi}{2})$$

apply Rule 5a

$$= -4(\frac{\pi}{2}, 0) + 4(0, \frac{\pi}{2})$$

$$= 4(0, \frac{\pi}{2}) + 4(0, \frac{\pi}{2})$$

$$= 8(0, \frac{\pi}{2})$$
(reuse point)

apply Rule 3a

$$= 8 \cdot \frac{1}{n} \sin \frac{n\pi}{2}$$

$$a_n\pi = \frac{8}{n} \sin \frac{n\pi}{2}$$

$$a_n = \frac{8}{n\pi} \sin \frac{n\pi}{2}$$

Therefore

$$a_n = \begin{cases} 0 & n \text{ is even} \\[2mm] \dfrac{8}{n\pi} & n = 1, 5, 9, \ldots \\[2mm] -\dfrac{8}{n\pi} & n = 3, 7, 11, \ldots \end{cases}$$

Solution for a_0   Continuing from the reuse point,

$$a_0\pi = 8(0, \frac{\pi}{2})$$

apply Rule 3c

$$= 8 \cdot \frac{\pi}{2}$$

$$a_0\pi = 4\pi$$

$$\boxed{a_0 = 4}$$

Solution for Fourier series

$$f(x) = \frac{a_0}{2} + \sum_{n=1}^{\infty} a_n \cos nx$$

$$= \frac{4}{2} + a_n(\cos x + \cos 5x + \cdots) + a_n(\cos 3x + \cos 7x + \cdots)$$

$$= 2 + \frac{8}{n\pi}(\cos x + \cos 5x + \cdots) - \frac{8}{n\pi}(\cos 3x + \cos 7x + \cdots)$$

$$= 2 + \frac{8}{\pi}(\cos x + \frac{1}{5}\cos 5x + \cdots) - \frac{8}{\pi}(\cos 3x + \frac{1}{7}\cos 7x + \cdots)$$

$$\boxed{f(x) = 2 + \frac{8}{\pi}(\cos x - \frac{1}{3}\cos 3x + \cdots)}$$

Example 2

Find the Fourier series for the function $f(x)$ given that the graph is as shown below. Assume period is 2π.

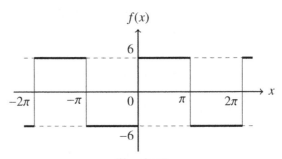

Fig. 6.17

Determine the property of $f(x)$ The function is an odd function, thus

$$\boxed{a_n = 0} \quad \text{and} \quad \boxed{a_0 = 0}$$

Solution for analytical definition   We can define the function analytically as:

$$f(x) = \begin{cases} -6 & -\pi < x < 0 \\ 6 & 0 < x < \pi \end{cases}$$

$$f(x + 2\pi) = f(x)$$

Solution for b_n

$$b_n \pi = \text{sum of cases}$$
$$= -6(-\pi, 0) + 6(0, \pi)$$

apply Rule 5b

$$= -6(\pi, 0) + 6(0, \pi)$$
$$= 6(0, \pi) + 6(0, \pi)$$
$$b_n \pi = 12(0, \pi)$$

apply Rule 3b

$$= 12 \cdot \frac{1}{n}(1 - \cos n\pi)$$
$$b_n = \frac{12}{n\pi}(1 - \cos n\pi)$$

Therefore

$$\boxed{b_n = \begin{cases} \dfrac{24}{n\pi} & n \text{ is odd} \\ 0 & n \text{ is even} \end{cases}}$$

Solution for Fourier series

$$f(x) = \sum_{n=1}^{\infty} b_n \sin nx$$

$$= b_n(\sin x + \sin 3x + \cdots) + b_n(\sin 2x + \sin 4x + \cdots)$$

$$= \frac{24}{n\pi}(\sin x + \sin 3x + \cdots) + 0(\sin 2x + \sin 4x + \cdots)$$

$$= \frac{24}{\pi}(\sin x + \frac{1}{3}\sin 3x + \cdots) + 0$$

$$\boxed{f(x) = \frac{24}{\pi}(\sin x + \frac{1}{3}\sin 3x + \cdots)}$$

Example 3

Find the Fourier series for the function $f(x)$ given that the graph is as shown below. Assume period is 2π.

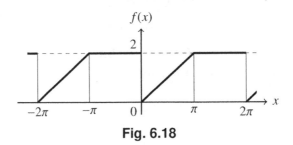

Fig. 6.18

Determine the property of $f(x)$ This is neither an odd nor an even function. Therefore we must solve for a_0, a_n and b_n.

Solution for analytical definition   We can define the function analytically as:

$$f(x) = \begin{cases} 2 & -\pi < x < 0 \\ \dfrac{2}{\pi}x & 0 < x < \pi \end{cases}$$

$$f(x + 2\pi) = f(x)$$

Solution for b_n

$$b_n \pi = \text{sum of cases}$$

$$= 2(-\pi, 0) + \frac{2}{\pi} x(0, \pi) \qquad \text{(reuse point 1)}$$

apply Rule 5b

$$= 2(\pi, 0) + \frac{2}{\pi} x(0, \pi)$$

$$= -2(0, \pi) + \frac{2}{\pi} x(0, \pi)$$

apply Rules 3b and 4b

$$= -\frac{2}{n}(1 - \cos n\pi) + \frac{2}{\pi}\left(-\frac{\pi}{n}\cos n\pi + \frac{1}{n^2}\sin n\pi\right)$$

$$= -\frac{2}{n} + \frac{2}{n}\cos n\pi - \frac{2}{n}\cos n\pi + 0$$

$$b_n \pi = -\frac{2}{n} + 0$$

$$\boxed{b_n = -\frac{2}{n\pi}}$$

Solution for a_n   Continuing from reuse point 1,

$$a_n \pi = 2(-\pi, 0) + \frac{2}{\pi} x(0, \pi)$$

apply Rule 5a

$$= -2(\pi, 0) + \frac{2}{\pi} x(0, \pi)$$

$$= 2(0, \pi) + \frac{2}{\pi} x(0, \pi) \qquad \text{(reuse point 2)}$$

apply Rules 3a and 4a

$$= \frac{2}{n}\sin n\pi + \frac{2}{\pi}\left[\frac{\pi}{n}\sin n\pi + \frac{1}{n^2}(\cos n\pi - 1)\right]$$

$$= 0 + \frac{2}{\pi}\left[0 + \frac{1}{n^2}(\cos n\pi - 1)\right]$$

$$a_n \pi = \frac{2}{\pi n^2}(\cos n\pi - 1)$$

$$a_n = \frac{2}{\pi^2 n^2}(\cos n\pi - 1)$$

Therefore

$$a_n = \begin{cases} -\dfrac{4}{\pi^2 n^2} & n \text{ is odd} \\[2mm] 0 & n \text{ is even} \end{cases}$$

Solution for a_0   Continuing from reuse point 2,

$$a_0 \pi = 2(0, \pi) + \frac{2}{\pi} x(0, \pi)$$

apply Rules 3c and 4c

$$= 2(\pi) + \frac{2}{\pi}(\frac{\pi^2}{2})$$

$$a_0 \pi = 2\pi + \pi$$

$$\boxed{a_0 = 3}$$

Solution for Fourier series

$$f(x) = \frac{a_0}{2} + \sum_{n=1}^{\infty} a_n \cos nx + \sum_{n=1}^{\infty} b_n \sin nx$$

$$= \frac{3}{2} + a_n(\cos x + \cos 2x + \cdots) + b_n(\sin x + \sin 2x + \cdots)$$

$$= \frac{3}{2} + a_n(\cos x + \cos 3x + \cdots) + a_n(\cos 2x + \cos 4x + \cdots)$$

$$+ b_n(\sin x + \sin 2x + \cdots)$$

$$= \frac{3}{2} - \frac{4}{\pi^2 n^2}(\cos x + \cos 3x + \cdots) + 0(\cos 2x + \cos 4x + \cdots)$$

$$- \frac{2}{n\pi}(\sin x + \sin 2x + \cdots)$$

$$\boxed{f(x) = \frac{3}{2} - \frac{4}{\pi^2}(\cos x + \frac{1}{3^2}\cos 3x + \cdots) - \frac{2}{\pi}(\sin x + \frac{1}{2}\sin 2x + \cdots)}$$

Exercise 3

Find the Fourier series for the following functions.

1.

$$f(x) = \begin{cases} -4x & -\pi < x < 0 \\ 4x & 0 < x < \pi \end{cases}$$

$$f(x + 2\pi) = f(x)$$

2.

$$f(x) = \begin{cases} -\dfrac{\pi}{4} & -\pi < x < 0 \\ \dfrac{\pi}{4} & 0 < x < \pi \end{cases}$$

$$f(x + 2\pi) = f(x)$$

3.

$$f(x) = \begin{cases} 2 & -2 < x < 0 \\ 0 & 0 < x < 2 \end{cases}$$

$$f(x + 2\pi) = f(x)$$

4.

$$f(x) = \begin{cases} x & -\dfrac{\pi}{2} < x < \dfrac{\pi}{2} \\ \pi - x & \dfrac{\pi}{2} < x < \dfrac{3\pi}{2} \end{cases}$$

$$f(x + 2\pi) = f(x)$$

5.

$$f(x) = \begin{cases} 1 - \dfrac{1}{2}|x| & -2 < x < 2 \\ 0 & 2 < x < 6 \end{cases}$$

$$f(x + 8) = f(x)$$

CHAPTER 7

Half-Range Series

We have discussed representing periodic functions as Fourier series. What about functions that are not periodic? Half-range series is a method of writing a non-periodic function as a Fourier series. The Fourier series produced in this method is special because it is made up entirely of sine or cosine functions. We have freedom to decide which it is for a function.

Consider the piecewise function $g(x)$ shown below. This function is not periodic. We wish to represent $g(x)$ by a Fourier series. We could do this in at least three ways.

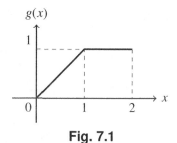

$g(x)$

Fig. 7.1

Case I   Here, we shall create a periodic function $f(x)$ by repeating the graph of $g(x)$ throughout the x-axis. Then we would find the Fourier series of $f(x)$. This Fourier series will then agree with $g(x)$ in the interval $0 < x < 2$. The graph of $f(x)$ is shown in Fig. 7.2.

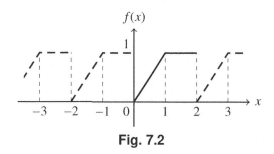

$f(x)$

Fig. 7.2

The problem with doing it this way is that we would need to compute all three Fourier coefficients a_0, a_n and b_n. This is not required in the other ways.

111

Case II   Define $f(x)$ such that it is an even periodic function. This will make the work easier because we would only need to find two instead of three Fourier coefficients. That is, a_0 and a_n. The graph of $f(x)$ is shown in Fig. 7.3.

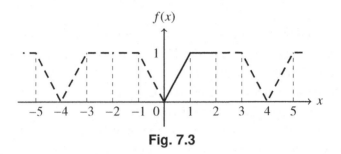

Fig. 7.3

Precisely, we have made $f(x)$ to satisfy the three conditions below.

• It is equal to $g(x)$ in the interval $0 < x < 2$. In math notation, this means:

$$f(x) = g(x) \quad 0 < x < 2$$

• It is an even periodic function with period 2π. In math notation, this means:

$$f(x) = f(-x) \quad \text{and} \quad f(x + 2\pi) = f(x) \quad \text{for all } x$$

When $f(x)$ is so defined, it is known as the *even periodic extension* of $g(x)$.

Case III   Define $f(x)$ such that it is an odd periodic function. Thus we only need to solve for b_n. The graph of $f(x)$ is shown in Fig. 7.4.

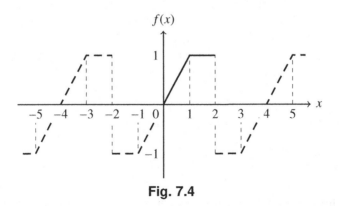

Fig. 7.4

Precisely, we have made $f(x)$ to satisfy the three conditions below.

- It is equal to $g(x)$ in the interval $0 < x < 2$. In math notation, this means:

$$f(x) = g(x) \quad 0 < x < 2$$

- It is an odd periodic function with period 2π. In math notation, this means:

$$f(x) = -f(-x) \quad \text{and} \quad f(x + 2\pi) = f(x) \quad \text{for all } x$$

When $f(x)$ is so defined, it is known as the *odd periodic extension* of $g(x)$.

Half-Range Series We only make use of Cases II and III in Half-range series. In Case II, $f(x)$ is even; thus the Fourier series will be made up entirely of cosine functions. This is called the *cosine series* of $g(x)$ (where $g(x)$ is the initial function required to be represented as a Fourier series).

In Case III, $f(x)$ is odd; thus the Fourier series will be made up entirely of sine functions. As expected, this is called the *sine series* of $g(x)$.

Notice that the range for which $g(x)$ is defined ($0 < x < 2$) is exactly "half" the period (4 units) of $f(x)$. Hence the Fourier series of $f(x)$ is called the *half-range expansion* of $g(x)$.

Example 1

Consider the non-periodic function $g(x)$ shown below. Find the even and odd periodic extensions of $g(x)$.

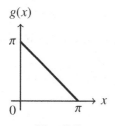

$$g(x) = -x + \pi \qquad 0 < x < \pi$$

Fig. 7.5

Even Periodic Extension   The even periodic extension is gotten in two steps. First we draw the even extension $f_e(x)$ of $g(x)$. See Fig. 7.6.

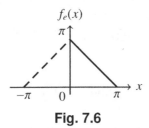

Fig. 7.6

The next step is to reproduce the graph above throughout the x-axis. See Fig. 7.7. Thus we have produced $f(x)$, the even periodic extension of $g(x)$.

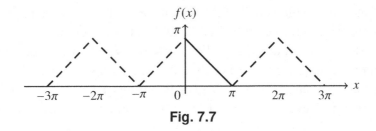

Fig. 7.7

From this graph, we can get the definition of $f(x)$ as:

$$f(x) = \begin{cases} x + \pi & -\pi < x < 0 \\ -x + \pi & 0 < x < \pi \end{cases}$$

$$f(x + 2\pi) = f(x)$$

Odd Periodic Extension   The odd periodic extension is also gotten in two steps. First we draw the odd extension $f_d(x)$ of $g(x)$. See Fig. 7.8.

The next step is to reproduce the graph above throughout the x-axis. See Fig. 7.9. Thus we have produced $f(x)$, the odd periodic extension of $g(x)$.

From this graph, we can get the definition of $f(x)$ as:

$$f(x) = \begin{cases} -x - \pi & -\pi < x < 0 \\ -x + \pi & 0 < x < \pi \end{cases}$$

$$f(x + 2\pi) = f(x)$$

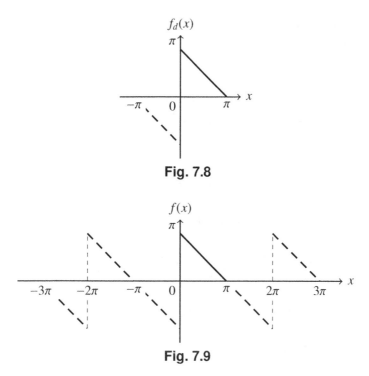

Fig. 7.8

Fig. 7.9

Example 2

Find a Fourier cosine series for the function $g(x)$ shown below.

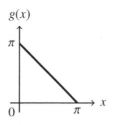

Fig. 7.10

Solution   In order to get a cosine series, we need to convert $g(x)$ to an even periodic function $f(x)$:

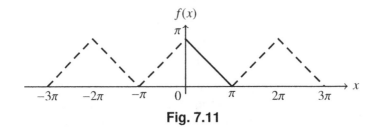

Fig. 7.11

From this graph, we can get the definition of $f(x)$ as:

$$f(x) = \begin{cases} x + \pi & -\pi < x < 0 \\ -x + \pi & 0 < x < \pi \end{cases}$$

$$f(x + 2\pi) = f(x)$$

Solution for ω   The period T of $f(x)$ is 2π. Thus

$$\omega = \frac{2\pi}{T}$$

$$= \frac{2\pi}{2\pi}$$

$$\boxed{\omega = 1}$$

Since we are looking for a cosine series, the Fourier series is made up of only cosine terms. Thus

$$\boxed{b_n = 0}$$

Solution for a_n

$$a_n \frac{\pi}{\omega} = \text{sum of cases of } f(x)$$

put $\omega = 1$

$$a_n \pi = (x + \pi)(-\pi, 0) + (-x + \pi)(0, \pi)$$
$$= x(-\pi, 0) + \pi(-\pi, 0) - x(0, \pi) + \pi(0, \pi)$$
$$= -x(0, -\pi) - \pi(0, -\pi) - x(0, \pi) + \pi(0, \pi)$$

apply Rules 5a and 6a

$$= -x(0, \pi) + \pi(0, \pi) - x(0, \pi) + \pi(0, \pi)$$

$$= -2x(0, \pi) + 2\pi(0, \pi) \qquad\qquad \text{(reuse point)}$$

$$= -2\left[-\frac{\pi}{n\omega} \sin n\omega\pi + \frac{1}{n^2\omega^2}(\cos n\omega\pi - 1) \right] + 2\pi\left[\frac{1}{n\omega} \sin n\omega\pi \right]$$

put $\omega = 1$

$$= -2\left[-\frac{\pi}{n} \sin n\pi + \frac{1}{n^2}(\cos n\pi - 1) \right] + 2\pi\left[\frac{1}{n} \sin n\pi \right]$$

$$a_n\pi = \left[0 - \frac{2}{n^2}(\cos n\pi - 1) \right] + 0$$

$$a_n = -\frac{2}{\pi n^2}(\cos n\pi - 1)$$

Thus

$$a_n = \begin{cases} \dfrac{4}{\pi n^2} & n \text{ is odd} \\ 0 & n \text{ is even} \end{cases}$$

Solution for a_0   Continuing from the reuse point,

$$a_0\pi = -2x(0, \pi) + 2\pi(0, \pi)$$

$$= -2 \cdot \frac{\pi^2}{2} + 2\pi(\pi)$$

$$= \pi^2$$

$$\boxed{a_0 = \pi}$$

Solution for cosine series

$$f(x) = \frac{a_0}{2} + \sum_{n=1}^{\infty} a_n \cos nx$$

$$= \frac{\pi}{2} + a_n(\cos x + \cos 3x + \cdots) + a_n(\cos 2x + \cos 4x + \cdots)$$

$$= \frac{\pi}{2} + \frac{4}{\pi n^2}(\cos x + \cos 3x + \cdots) + 0(\cos 2x + \cos 4x + \cdots)$$

$$\boxed{f(x) = \frac{\pi}{2} + \frac{4}{\pi}(\cos x + \frac{1}{3^2} \cos 3x + \cdots)}$$

This is the cosine series expansion of $g(x)$

Example 3

Find a Fourier sine series for the function $g(x)$ shown below.

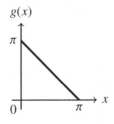

$g(x)$

π

0 π x

Fig. 7.12

Solution   In order to get a sine series, we need to convert $g(x)$ to an odd periodic function $f(x)$. See Fig. 7.13.

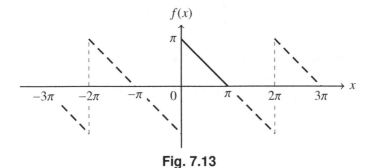

$f(x)$

π

-3π -2π $-\pi$ 0 π 2π 3π x

Fig. 7.13

From this graph, we can get the definition of $f(x)$ as:

$$f(x) = \begin{cases} -x - \pi & -\pi < x < 0 \\ -x + \pi & 0 < x < \pi \end{cases}$$

$$f(x + 2\pi) = f(x)$$

Solution for ω The period T of $f(x)$ is 2π. Thus

$$\omega = \frac{2\pi}{T}$$

$$= \frac{2\pi}{2\pi}$$

$$\boxed{\omega = 1}$$

Since we are looking for a sine series, the Fourier series is made up of only sine terms. Thus

$$\boxed{a_0 = 0} \quad \text{and} \quad \boxed{a_n = 0}$$

Solution for b_n

$b_n \dfrac{\pi}{\omega} = $ sum of cases of $f(x)$

put $\omega = 1$

$b_n \pi = (-x - \pi)(-\pi, 0) + (-x + \pi)(0, \pi)$

$\quad = -x(-\pi, 0) - \pi(-\pi, 0) - x(0, \pi) + \pi(0, \pi)$

$\quad = x(0, -\pi) + \pi(0, -\pi) - x(0, \pi) + \pi(0, \pi)$

apply Rules 5b and 6b

$\quad = -x(0, \pi) + \pi(0, \pi) - x(0, \pi) + \pi(0, \pi)$

$\quad = -2x(0, \pi) + 2\pi(0, \pi)$

$\quad = -2\left[-\dfrac{\pi}{n\omega} \cos n\omega\pi + \dfrac{1}{n^2\omega^2} \sin n\omega\pi \right] + 2\pi\left[\dfrac{1}{n\omega}(1 - \cos n\omega\pi) \right]$

put $\omega = 1$

$\quad = -2\left[-\dfrac{\pi}{n} \cos n\pi + \dfrac{1}{n^2} \sin n\pi \right] + 2\pi\left[\dfrac{1}{n}(1 - \cos n\pi) \right]$

$\quad = \left[\dfrac{2\pi}{n} \cos n\pi + 0 \right] + \dfrac{2\pi}{n}(1 - \cos n\pi)$

$b_n \pi = \dfrac{2\pi}{n}(\cos n\pi + 1 - \cos n\pi)$

$$\boxed{b_n = \dfrac{2}{n}}$$

Solution for sine series

$$f(x) = \sum_{n=1}^{\infty} b_n \sin nx$$

$$= b_n(\sin x + \sin 2x + \cdots)$$

$$= \frac{2}{n}(\sin x + \sin 2x + \cdots)$$

$$\boxed{f(x) = 2(\sin x + \frac{1}{2}\sin 2x + \cdots)}$$

This is the sine series expansion of $g(x)$

7.1   The Shortcut Method

There is a quicker way to find the sine or cosine series of a function $g(x)$. This is the shortcut method. It does not require drawing any graphs; instead we use the analytical definition of $g(x)$.

The following examples illustrate this method. The first two examples are the same with the previous two, except that here the shortcut method is used.

Example 1

Find a Fourier cosine series for the function $g(x)$ shown below.

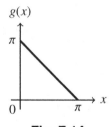

Fig. 7.14

Solution for analytical definition The analytical definition of $g(x)$ is

$$g(x) = -x + \pi \qquad 0 < x < \pi$$

Solution for ω The function $g(x)$ is defined in the range $0 < x < \boxed{\pi}$. Thus π is the upper bound of this range. The period T of $f(x)$ will be twice the upper bound of $g(x)$. That is,

$$T = 2 \times \pi = 2\pi$$

Thus

$$\omega = \frac{2\pi}{T}$$
$$= \frac{2\pi}{2\pi}$$
$$\boxed{\omega = 1}$$

If the function $g(x)$ was defined in the range $0 < x < \boxed{5}$. Then the period T will be $T = 2 \times 5 = 10$

Solution for a_n

$$a_n \frac{\pi}{\omega} = 2 \times \text{sum of cases of } g(x)$$

put $\omega = 1$

$$a_n \pi = 2(-x + \pi)(0, \pi)$$
$$= -2x(0, \pi) + 2\pi(0, \pi) \qquad \text{(reuse point)}$$
$$= -2\left[-\frac{\pi}{n\omega} \sin n\omega\pi + \frac{1}{n^2\omega^2}(\cos n\omega\pi - 1) \right] + 2\pi\left[\frac{1}{n\omega} \sin n\omega\pi \right]$$

put $\omega = 1$

$$= -2\left[-\frac{\pi}{n} \sin n\pi + \frac{1}{n^2}(\cos n\pi - 1) \right] + 2\pi\left[\frac{1}{n} \sin n\pi \right]$$
$$a_n \pi = \left[0 - \frac{2}{n^2}(\cos n\pi - 1) \right] + 0$$
$$a_n = -\frac{2}{\pi n^2}(\cos n\pi - 1)$$

Thus

$$a_n = \begin{cases} \dfrac{4}{\pi n^2} & n \text{ is odd} \\ 0 & n \text{ is even} \end{cases}$$

Solution for a_0   Continuing from the reuse point,

$$a_0\pi = -2x(0, \pi) + 2\pi(0, \pi)$$

$$= -2 \cdot \frac{\pi^2}{2} + 2\pi(\pi)$$

$$= \pi^2$$

$$\boxed{a_0 = \pi}$$

Solution for cosine series

$$f(x) = \frac{a_0}{2} + \sum_{n=1}^{\infty} a_n \cos nx$$

$$= \frac{\pi}{2} + a_n(\cos x + \cos 3x + \cdots) + a_n(\cos 2x + \cos 4x + \cdots)$$

$$= \frac{\pi}{2} + \frac{4}{\pi n^2}(\cos x + \cos 3x + \cdots) + 0(\cos 2x + \cos 4x + \cdots)$$

$$\boxed{f(x) = \frac{\pi}{2} + \frac{4}{\pi}(\cos x + \frac{1}{3^2} \cos 3x + \cdots)}$$

This is the cosine series expansion of $g(x)$

Example 2

Find a Fourier sine series for the function $g(x)$ shown below.

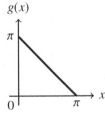

Fig. 7.15

Solution for analytical definition   The analytical definition of $g(x)$ is

$$g(x) = -x + \pi \qquad 0 < x < \pi$$

Solution for ω   The period T of $f(x)$ will be twice the upper bound of $g(x)$. That is,

$$T = 2 \times \pi = 2\pi$$

Thus

$$\omega = \frac{2\pi}{T}$$
$$= \frac{2\pi}{2\pi}$$
$$\boxed{\omega = 1}$$

Solution for b_n

$$b_n \frac{\pi}{\omega} = 2 \times \text{sum of cases of } g(x)$$

put $\omega = 1$

$$b_n \pi = 2(-x + \pi)(0, \pi)$$
$$= -2x(0, \pi) + 2\pi(0, \pi)$$
$$= -2\left[-\frac{\pi}{n\omega} \cos n\omega\pi + \frac{1}{n^2\omega^2} \sin n\omega\pi \right] + 2\pi\left[\frac{1}{n\omega}(1 - \cos n\omega\pi) \right]$$

put $\omega = 1$

$$= -2\left[-\frac{\pi}{n} \cos n\pi + \frac{1}{n^2} \sin n\pi \right] + 2\pi\left[\frac{1}{n}(1 - \cos n\pi) \right]$$
$$= \left[\frac{2\pi}{n} \cos n\pi + 0 \right] + \frac{2\pi}{n}(1 - \cos n\pi)$$
$$b_n \pi = \frac{2\pi}{n}(\cos n\pi + 1 - \cos n\pi)$$
$$\boxed{b_n = \frac{2}{n}}$$

Solution for sine series

$$f(x) = \sum_{n=1}^{\infty} b_n \sin nx$$

$$= b_n(\sin x + \sin 2x + \cdots)$$

$$= \frac{2}{n}(\sin x + \sin 2x + \cdots)$$

$$\boxed{f(x) = 2(\sin x + \frac{1}{2}\sin 2x + \cdots)}$$

This is the sine series expansion of $g(x)$

Example 3

Determine the sine series to represent the function defined as

$$g(x) = \begin{cases} 1 & 0 < x < 1 \\ 2 & 1 < x < 2 \end{cases}$$

Solution for ω   The function $g(x)$ is defined in the range $0 < x < \boxed{2}$. Thus 2 is the upper bound of this range. The period T of $f(x)$ will be twice the upper bound of $g(x)$. That is,

$$T = 2 \times 2 = 4$$

Thus

$$\omega = \frac{2\pi}{T}$$

$$= \frac{2\pi}{4}$$

$$\boxed{\omega = \frac{\pi}{2}}$$

Since we are looking for a sine series, the Fourier series is made up of only sine terms. Thus

$$\boxed{a_0 = 0} \quad \text{and} \quad \boxed{a_n = 0}$$

Solution for b_n

$$b_n \frac{\pi}{\omega} = 2 \times \text{sum of cases of } g(x)$$

$$\text{put } \omega = \frac{\pi}{2}$$

$$2b_n = 2 \times \text{sum of cases of } g(x)$$

$$b_n = \text{sum of cases of } g(x)$$

$$= 1(0, 1) + 2(1, 2)$$

$$= (0, 1) + 2(1, 0) + 2(0, 2)$$

$$= (0, 1) - 2(0, 1) + 2(0, 2)$$

$$= -(0, 1) + 2(0, 2)$$

$$= -\frac{1}{n\omega}[1 - \cos n\omega(1)] + 2[\frac{1}{n\omega}(1 - \cos n\omega(2))]$$

$$= -\frac{1}{n\omega}(1 - \cos n\omega) + \frac{2}{n\omega}(1 - \cos 2n\omega)$$

$$\text{put } \omega = \frac{\pi}{2}$$

$$= -\frac{1}{n(\pi/2)}(1 - \cos \frac{n\pi}{2}) + \frac{2}{n(\pi/2)}(1 - \cos n\pi)$$

$$= -\frac{2}{n\pi}(1 - \cos \frac{n\pi}{2}) + \frac{4}{n\pi}(1 - \cos n\pi)$$

$$= \frac{2}{n\pi}[-(1 - \cos \frac{n\pi}{2}) + 2(1 - \cos n\pi)]$$

$$= \frac{2}{n\pi}[-1 + \cos \frac{n\pi}{2} + 2 - 2\cos n\pi]$$

$$b_n = \frac{2}{n\pi}[\cos \frac{n\pi}{2} - 2\cos n\pi + 1]$$

Recall that

$$\cos \frac{n\pi}{2} = \begin{cases} 0 & n \text{ is odd} \\ -1 & n = 2, 6, 10, \ldots \\ 1 & n = 4, 8, 12, \ldots \end{cases}$$

and

$$\cos n\pi = \begin{cases} -1 & n \text{ is odd} \\ 1 & n \text{ is even} \end{cases}$$

When n is odd $\cos \dfrac{n\pi}{2} = 0$ and $\cos n\pi = -1$

Thus,

$$b_n = \frac{2}{n\pi}[0 - 2(-1) + 1]$$

$$= \frac{2}{n\pi}(3)$$

$$= \frac{6}{n\pi}$$

When $n = 2, 6, 10, \dots$ $\cos \dfrac{n\pi}{2} = -1$ and $\cos n\pi = 1$

Thus,

$$b_n = \frac{2}{n\pi}(-1 - 2(1) + 1)$$

$$= \frac{2}{n\pi}(-2)$$

$$= -\frac{4}{n\pi}$$

When $n = 4, 8, 12, \dots$ $\cos \dfrac{n\pi}{2} = 1$ and $\cos n\pi = 1$

Thus,

$$b_n = \frac{2}{n\pi}(1 - 2(1) + 1)$$

$$= \frac{2}{n\pi}(0)$$

$$= 0$$

Therefore,

$$b_n = \begin{cases} \dfrac{6}{n\pi} & n \text{ is odd} \\[2mm] -\dfrac{4}{n\pi} & n = 2, 6, 10, \dots \\[2mm] 0 & n = 4, 8, 12, \dots \end{cases}$$

Solution for Fourier series

$$f(x) = \sum_{n=1}^{\infty} b_n \sin n\omega x$$

$$= b_n(\sin \frac{\pi x}{2} + \sin 3\frac{\pi x}{2} + \cdots) + b_n(\sin 2\frac{\pi x}{2} + \sin 6\frac{\pi x}{2} + \cdots)$$

$$+ b_n(\sin 4\frac{\pi x}{2} + \sin 8\frac{\pi x}{2} + \cdots)$$

$$= \frac{6}{n\pi}(\sin \frac{\pi x}{2} + \sin 3\frac{\pi x}{2} + \cdots) - \frac{4}{n\pi}(\sin 2\frac{\pi x}{2} + \sin 6\frac{\pi x}{2} + \cdots)$$

$$+ 0(\sin 4\frac{\pi x}{2} + \sin 8\frac{\pi x}{2} + \cdots)$$

$$\boxed{f(x) = \frac{6}{\pi}(\sin \frac{\pi x}{2} + \frac{1}{3} \sin 3\frac{\pi x}{2} + \cdots) - \frac{4}{\pi}(\frac{1}{2} \sin 2\frac{\pi x}{2} + \frac{1}{6} \sin 6\frac{\pi x}{2} + \cdots)}$$

Exercise

Find

   (a) the Fourier cosine series

   (b) the Fourier sine series

of the following functions.

1.

$$g(x) = 1 \qquad 0 < x < 2$$

2.

$$g(x) = x \qquad 0 < x < \frac{1}{2}$$

3.

$$g(x) = 4 \qquad 0 < x < 3$$

4.

$$g(x) = \begin{cases} 0 & 0 < x < 2 \\ 1 & 2 < x < 4 \end{cases}$$

5.

$$g(x) = \begin{cases} 1 & 0 < x < 1 \\ 2 & 1 < x < 2 \end{cases}$$

6.

$$g(x) = \begin{cases} 0 & 0 < x < \frac{\pi}{2} \\ 1 & \frac{\pi}{2} < x < \pi \end{cases}$$

7.

$$g(x) = \begin{cases} x & 0 < x < \frac{\pi}{2} \\ \pi - x & \frac{\pi}{2} < x < \pi \end{cases}$$

8.

$$g(x) = \begin{cases} \dfrac{1}{2} & 0 < x < \dfrac{5}{2} \\[2mm] 1 & \dfrac{5}{2} < x < 5 \end{cases}$$

9.

$$g(x) = \begin{cases} x & 0 < x < \dfrac{1}{2} \\[2mm] 1 - x & \dfrac{1}{2} < x < 1 \end{cases}$$

10.

$$g(x) = \begin{cases} 1 & 0 < x < 1 \\ 0 & 1 < x < 3 \\ -1 & 3 < x < 5 \end{cases}$$

CHAPTER 8

Fourier Series and Points

This chapter discusses how Fourier series behaves at certain points. We investigate how jump discontinuities affect Fourier series and the value of a Fourier series at "special" points.

What is the value of a Fourier series at a point? What does "point" even mean in this context? The answer to these questions is discussed here.

"Point" in this context means a point in the function. See Fig. 8.1. The general way of representing a point in a function $f(x)$ is by coordinates: $(a, f(a))$. Thus we can refer to a point in a function like so: $(3, 5)$, which means $f(3) = 5$. However, each value of x corresponds to *only one* point in the function $f(x)$. Thus we can refer to a point in a function by the value of x alone. For example, instead of saying "point $(3, 5)$", we can just say "point $x = 3$".

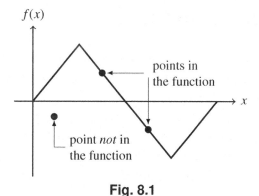

Fig. 8.1

We get the value of a Fourier series at a point by substituting the value of x into the Fourier series. An example should make this idea clear.

Example

Consider the Fourier series

$$f(x) = \sin x + \frac{1}{3}\sin 3x + \cdots$$

$$\text{put } x = 0$$

$$= \sin 0 + \frac{1}{3}\sin 0 + \cdots$$

$$= 0 + 0 + \cdots$$

$$\therefore f(0) = 0$$

Thus we say that the value of the Fourier series at the point $x = 0$ is zero.

8.1 Jump Discontinuity

What is a discontinuity? It is an abrupt change in the value of a function. A function that has a discontinuity is called *discontinuous*. Otherwise, it is *continuous*.

A jump discontinuity is one where the function jumps from one value to another at the same value of x. The graph of the function shows that the function stops at one place and picks up again at somewhere else.

Let's see an example of a function that has a jump discontinuity.

Example 1

Consider the function in Fig. 8.2.

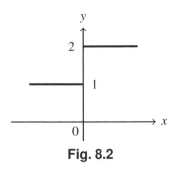

Fig. 8.2

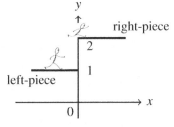

Fig. 8.3

Imagine you are walking from the left-piece to the right-piece. You will have to jump at $x = 0$ in order to get to the right-piece. See Fig. 8.3. Therefore, we say the function has a jump discontinuity at $x = 0$.

Example 2

As another example, consider the graph below

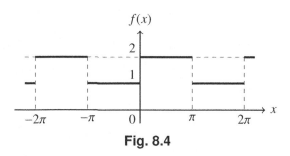

Fig. 8.4

The graph breaks at the points $x = -2\pi$, $-\pi$, 0, π and 2π. These are all jump discontinuities.

8.1.1   What is *not* a Discontinuity?

Just to be clear, we shall give examples of points that are sometimes mistaken to be discontinuous.

Example 3

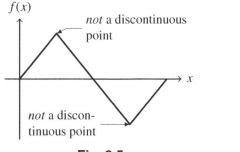

Fig. 8.5

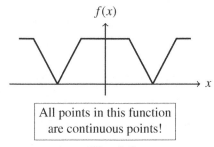

Fig. 8.6

8.2   Partial Sum of Fourier Series

The complete representation of $f(x)$ in Fourier series requires an infinite number of terms. However we can get a good approximation of $f(x)$ by stopping the summation of the Fourier series after a finite number of terms.

Consider the infinite series

$$\frac{1}{1^2} + \frac{1}{2^2} + \frac{1}{3^2} + \cdots$$

This series converges to the value $\frac{\pi^2}{6}$. We shall take partial sums of this series and observe how they approximate $\frac{\pi^2}{6}$. Let S_k be the partial sum of the series up to the kth term.

$$S_k = \frac{1}{1^2} + \frac{1}{2^2} + \frac{1}{3^2} + \cdots + \frac{1}{k^2}$$

In the following table, we shall observe how S_k gets closer to $\frac{\pi^2}{6}$ as we take more and more terms into the partial sum of the Fourier series.

Table 8.1

$$\frac{\pi^2}{6} = 1.6449 \text{ (to 4 d.p)}$$

k	S_k
1	1.0000
10	1.5498
100	1.6350
1000	1.6439
2000	1.6444

Just as the partial sums of this infinite series get closer to $\frac{\pi^2}{6}$ as we add more terms, so also the partial sums of the Fourier series of a function $f(x)$ gets closer to $f(x)$ as we add more terms.

Consider the Fourier series

$$\frac{\pi}{2} - (\sin x + \frac{1}{2}\sin 2x + \frac{1}{3}\sin 3x + \cdots)$$

The above Fourier series converges to the graph shown below.

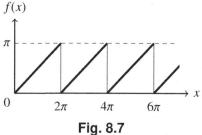

Fig. 8.7

We shall take partial sums of the Fourier series and observe how they approximate $f(x)$. Let S_k be the partial sum of the Fourier series up to the kth term.

$$S_k = \frac{\pi}{2} - \left(\sin x + \frac{1}{2} \sin 2x + \frac{1}{3} \sin 3x + \cdots + \frac{1}{k} \sin kx \right)$$

Table 8.2

k	graph of S_k	k	graph of S_k
1		10	
3		14	
5			

In Table 8.2, we observed how the graph of S_k gets closer to the graph of $f(x)$ as we add more and more terms into the partial sum. Thus we can conclude that the partial sum of the Fourier series approximates the function better as more terms are added.

8.3   Gibbs Phenomenon

The Gibbs phenomenon is the peculiar manner in which the Fourier series of a function behaves at a jump discontinuity.

From our previous set of graph diagrams, the graph of S_k at $k = 14$ is shown in Fig 8.8.

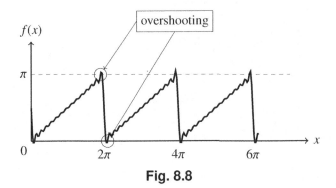

Fig. 8.8

Observe the spikes at the points of discontinuity, $x = 2\pi, 4\pi, 6\pi$, etc. These spikes in the graph are called overshooting. These overshooting are not present in the graph of $f(x)$ but in the graph of S_k. Thus they should be seen as defects in our approximation of $f(x)$. They do not die out as more and more terms are added to the sum. This phenomenon is known as the *Gibbs phenomenon*.

In summary,the Gibbs phenomenon involves two facts:

- the Fourier series overshoot at a jump discontinuity,

- this overshooting does not die out as more terms are added to the sum.

8.4 Sum of a Fourier Series at a Jump Discontinuity

A function $f(x)$ and its Fourier series may not be equal at all points in the function. Consider the function $f(x)$ below

$$f(x) = \begin{cases} -2 & -\pi < x < 0 \\ 2 & 0 \leq x < \pi \end{cases}$$

The Fourier series of $f(x)$ is

$$\frac{8}{\pi}(\sin x + \frac{1}{3}\sin 3x + \cdots)$$

Let us denote the Fourier series of $f(x)$ by $S(x)$. Thus

$$S(x) = \frac{8}{\pi}(\sin x + \frac{1}{3}\sin 3x + \cdots)$$

From the definition of $f(x)$, at $x = 0$, $f(x) = 2$. Now put $x = 0$ into $S(x)$.

$$S(0) = \frac{8}{\pi}(\sin 0 + \frac{1}{3}\sin 0 + \cdots)$$
$$= 0$$

Thus we have $f(0) = 2$ but $S(0) = 0$. Thus

$$f(0) \neq S(0)$$

But $S(x)$ is the Fourier series of $f(x)$. Should they not be equal? Since $f(0) \neq S(0)$, it means our Fourier series $S(x)$ is not actually equal to $f(x)$, at least not at every point.

Why did this happen? This happened because $x = 0$ is a point of discontinuity. We can observe this from the graph of $f(x)$ shown in Fig. 8.9.

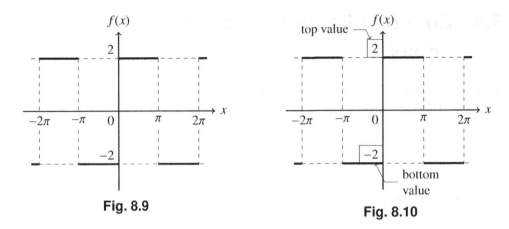

Fig. 8.9   **Fig. 8.10**

At a jump discontinuity, the value of $S(x)$ is the average of the "top" and "bottom" values of the jump. See Fig. 8.10.

From the graph,

$$\text{top value} = 2$$
$$\text{bottom value} = -2$$

Thus $S(0)$ is the average of 2 and -2.

$$
\begin{aligned}
S(0) &= \frac{2 + (-2)}{2} \\
&= \frac{2 - 2}{2} \\
&= 0
\end{aligned}
$$

Therefore, at a point of jump discontinuity, the Fourier series of a function converges to the average of the top and bottom values of the jump.

Example 1

Consider the function shown in Fig. 8.11.

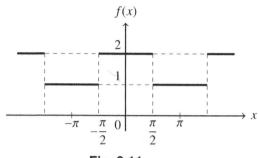

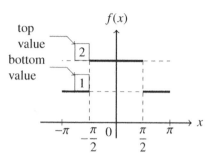

Fig. 8.11   **Fig. 8.12**

A jump discontinuity occurs at $x = -\dfrac{\pi}{2}$. What value does the Fourier series converge to at this point.

Solution   We can redraw the graph as shown in Fig. 8.12.
From the graph,

$$\text{top value} = 2$$
$$\text{bottom value} = 1$$

The value of the Fourier series $S(x)$ at $x = -\dfrac{\pi}{2}$ is denoted as $S(-\dfrac{\pi}{2})$. Thus $S(-\dfrac{\pi}{2})$ is the average of 2 and 1.

$$S(-\frac{\pi}{2}) = \frac{2+1}{2}$$

$$\boxed{S(-\frac{\pi}{2}) = \frac{3}{2}}$$

Therefore the Fourier series of $f(x)$ converges to $\dfrac{3}{2}$ at $x = -\dfrac{\pi}{2}$.

Example 2

Consider the function in Fig. 8.13.

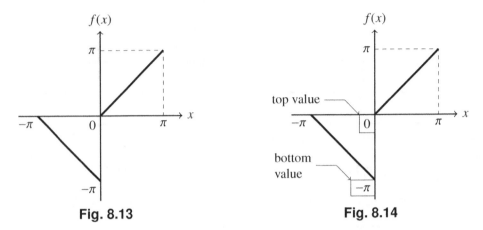

Fig. 8.13   **Fig. 8.14**

A jump discontinuity occurs at $x = 0$. What value does the Fourier series converge to at this point.

Solution   We can redraw the graph as shown in Fig. 8.14.
From the graph,

$$\text{top value} = 0$$
$$\text{bottom value} = -\pi$$

Thus $S(0)$ is the average of 0 and $-\pi$.

$$S(0) = \frac{0 + (-\pi)}{2}$$
$$S(0) = \frac{0 - \pi}{2}$$
$$\boxed{S(0) = -\frac{\pi}{2}}$$

Therefore the Fourier series of $f(x)$ converges to $-\dfrac{\pi}{2}$ at $x = 0$.

8.4.1   Sum of a Fourier series at a Point of Continuity

If a point is not discontinuous, then it is continuous. Points of continuity are "good" points. This is because the function $f(x)$ and its Fourier series $S(x)$ are equal at a point of continuity. Therefore, if $x = x_0$ is a point of continuity, then

$$f(x_0) = S(x_0)$$

8.5   Fourier Series at a Point of a Function

By evaluating the value of a Fourier series at a point of a function, we can derive some interesting results about the mathematical constant π. We can also use this method to find the value of very important infinite sums.

Example 1

Consider the function in Fig. 8.15.

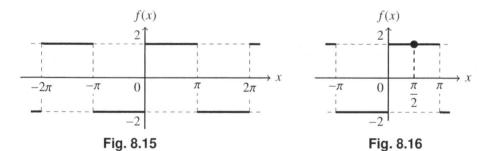

Fig. 8.15   Fig. 8.16

The Fourier series for $f(x)$ is

$$f(x) = \frac{8}{\pi}\left(\sin x + \frac{1}{3}\sin 3x + \cdots\right) \tag{8.1}$$

From the graph, let us determine the value of $f(x)$ at $x = \frac{\pi}{2}$. This is shown in Fig. 8.16.

At $x = \frac{\pi}{2}$, $f(x) = 2$

$$f(\frac{\pi}{2}) = 2 \tag{8.2}$$

Now put (8.2) in (8.1)

Therefore

$$2 = \frac{8}{\pi}[\sin\left(\frac{\pi}{2}\right) + \frac{1}{3}\sin 3\left(\frac{\pi}{2}\right) + \frac{1}{5}\sin 5\left(\frac{\pi}{2}\right) + \cdots]$$

$$\frac{2\pi}{8} = \sin\left(\frac{\pi}{2}\right) + \frac{1}{3}\sin 3\left(\frac{\pi}{2}\right) + \frac{1}{5}\sin 5\left(\frac{\pi}{2}\right) + \cdots$$

$$\frac{\pi}{4} = \sin\frac{\pi}{2} + \frac{1}{3}\sin\frac{3\pi}{2} + \frac{1}{5}\sin\frac{5\pi}{2} + \cdots \qquad (8.3)$$

Recall that

$$\sin\frac{n\pi}{2} = \begin{cases} 1 & n = 1, 5, 9, \ldots \\ -1 & n = 3, 7, 11, \ldots \end{cases}$$

Thus

$$\frac{\pi}{4} = 1 + \frac{1}{3}(-1) + \frac{1}{5}(1) + \frac{1}{7}(-1) + \cdots$$

$$\boxed{\frac{\pi}{4} = 1 - \frac{1}{3} + \frac{1}{5} - \frac{1}{7} + \cdots} \qquad (8.4)$$

Thus we have proved that

1. $\pi = 4\left(1 - \frac{1}{3} + \frac{1}{5} - \frac{1}{7} + \cdots\right)$

This is pi expressed as an infinite series.

2. $1 - \frac{1}{3} + \frac{1}{5} - \frac{1}{7} + \cdots = \frac{\pi}{4}$

We have been able to assign a value to the infinite series above.

Overview of the Steps

It all happened so fast. How did we arrive at the beautiful result in (8.4)? Let us see an overview of all the steps we carried out.

Step 1 We were given the graph of $f(x)$ and its Fourier series. From the graph of $f(x)$ we evaluated the value of $f(x)$ at $x = \frac{\pi}{2}$. We got $f(\frac{\pi}{2}) = 2$.

Step 2 We put $x = \frac{\pi}{2}$ and $f(x) = 2$ in the Fourier series and simplified the resulting equation. Then we got

$$\frac{\pi}{4} = 1 - \frac{1}{3} + \frac{1}{5} - \frac{1}{7} + \cdots$$

Example 2

Consider the function

$$f(x) = \begin{cases} -x - \pi & -\pi < x < 0 \\ x + \pi & 0 < x < \pi \end{cases}$$

$$f(x + 2\pi) = f(x)$$

The Fourier series for $f(x)$ is;

$$f(x) = \frac{\pi}{2} - \frac{4}{\pi}\left(\cos x + \frac{1}{3^2}\cos 3x + \cdots\right) + 4\left(\sin x + \frac{1}{3}\sin 3x + \cdots\right)$$

Derive π-related formulas from the Fourier series.

Solution   First, we shall draw the graph of $f(x)$

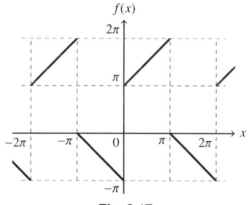

Fig. 8.17

Now look closely at the Fourier series of the function.

$$f(x) = \frac{\pi}{2} - \frac{4}{\pi}\left(\cos x + \frac{1}{3^2}\cos 3x + \cdots\right) + 4\left(\sin x + \frac{1}{3}\sin 3x + \cdots\right)$$

If we put $x = 0$ or $x = \pi$, all the sine terms will be zero. However, if we put $x = \frac{\pi}{2}$ all the cosine terms will be zero. Below, we shall try all three options.

Point $x = 0$ At $x = 0$, the graph shows a jump discontinuity.

$$\text{top value} = \pi$$
$$\text{bottom value} = -\pi$$
$$\text{average} = \frac{\pi + (-\pi)}{2}$$
$$= 0$$

Therefore, at $x = 0$, $f(x) = 0$. If we put $x = 0$, $f(x) = 0$ into the Fourier series, we get

$$0 = \frac{\pi}{2} - \frac{4}{\pi}\left(\cos 0 + \frac{1}{3^2}\cos 0 + \cdots\right) + 4\left(\sin 0 + \frac{1}{3}\sin 0 + \cdots\right)$$

$$0 = \frac{\pi}{2} - \frac{4}{\pi}\left(1 + \frac{1}{3^2}(1) + \cdots\right) + 4\left(0 + 0 + \cdots\right)$$

$$0 = \frac{\pi}{2} - \frac{4}{\pi}\left(1 + \frac{1}{3^2} + \frac{1}{5^2} + \cdots\right)$$

$$-\frac{\pi}{2} = -\frac{4}{\pi}\left(1 + \frac{1}{3^2} + \frac{1}{5^2} + \cdots\right)$$

$$\boxed{\frac{\pi^2}{8} = 1 + \frac{1}{3^2} + \frac{1}{5^2} + \frac{1}{7^2} + \cdots}$$

Point $x = \pi$ At $x = \pi$, the graph shows a jump discontinuity.

$$\text{top value} = 2\pi$$
$$\text{bottom value} = 0$$
$$\text{average} = \frac{2\pi + 0}{2}$$
$$= \pi$$

Therefore, at $x = \pi$, $f(x) = \pi$. If we put $x = \pi$, $f(x) = \pi$ into the Fourier series, we get

$$\pi = \frac{\pi}{2} - \frac{4}{\pi}\left(\cos \pi + \frac{1}{3^2}\cos 3\pi + \cdots\right) + 4\left(\sin \pi + \frac{1}{3}\sin 3\pi + \cdots\right)$$

$$\pi - \frac{\pi}{2} = -\frac{4}{\pi}\left(\cos \pi + \frac{1}{3^2}\cos 3\pi + \cdots\right) + 4\left(0 + 0 + \cdots\right)$$

$$\frac{\pi}{2} = -\frac{4}{\pi}\left(\cos \pi + \frac{1}{3^2}\cos 3\pi + \cdots\right) + 0$$

$$-\frac{\pi^2}{8} = \cos \pi + \frac{1}{3^2} \cos 3\pi + \frac{1}{5^2} \cos 5\pi + \frac{1}{7^2} \cos 7\pi + \cdots$$

Recall that $\cos n\pi = -1$ when n is odd. Therefore

$$-\frac{\pi^2}{8} = (-1) + \frac{1}{3^2}(-1) + \frac{1}{5^2}(-1) + \frac{1}{7^2}(-1) + \cdots$$

$$-\frac{\pi^2}{8} = (-1)\left(1 + \frac{1}{3^2} + \frac{1}{5^2} + \frac{1}{7^2} + \cdots\right)$$

$$\boxed{\frac{\pi^2}{8} = 1 + \frac{1}{3^2} + \frac{1}{5^2} + \frac{1}{7^2} + \cdots}$$

This is the same result we got at point $x = 0$.

Point $x = \dfrac{\pi}{2}$ At $x = \dfrac{\pi}{2}$, the graph shows a point of continuity. According to the piece-wise definition we have that

$$f(x) = x + \pi \qquad 0 < x < \pi$$

Therefore

$$f(\frac{\pi}{2}) = \frac{\pi}{2} + \pi$$

$$= \frac{3\pi}{2}$$

Put $x = \dfrac{\pi}{2}$, $f(x) = \dfrac{3\pi}{2}$ into the Fourier series.

$$\frac{3\pi}{2} = \frac{\pi}{2} - \frac{4}{\pi}\left(\cos\left(\frac{\pi}{2}\right) + \frac{1}{3^2} \cos 3\left(\frac{\pi}{2}\right) + \cdots\right) + 4\left(\sin\left(\frac{\pi}{2}\right) + \frac{1}{3} \sin 3\left(\frac{\pi}{2}\right) + \cdots\right)$$

$$\frac{3\pi}{2} - \frac{\pi}{2} = -\frac{4}{\pi}\left(\cos \frac{\pi}{2} + \frac{1}{3^2} \cos \frac{3\pi}{2} + \cdots\right) + 4\left(\sin \frac{\pi}{2} + \frac{1}{3} \sin \frac{3\pi}{2} + \cdots\right)$$

$$\pi = -\frac{4}{\pi}\left(\cos \frac{\pi}{2} + \frac{1}{3^2} \cos \frac{3\pi}{2} + \cdots\right) + 4\left(\sin \frac{\pi}{2} + \frac{1}{3} \sin \frac{3\pi}{2} + \cdots\right)$$

Recall that

$$\cos \frac{n\pi}{2} = 0 \quad (n \text{ is odd}) \qquad \text{and} \qquad \sin \frac{n\pi}{2} = \begin{cases} 1 & n = 1, 5, 9, \ldots \\ -1 & n = 3, 7, 11, \ldots \end{cases}$$

Thus

$$\pi = -\frac{4}{\pi}\left(0 + 0 + \cdots\right) + 4\left(1 + \frac{1}{3}(-1) + \frac{1}{5}(1) + \frac{1}{7}(-1) + \cdots\right)$$

$$= 0 + 4\left(1 + \frac{1}{3}(-1) + \frac{1}{5}(1) + \frac{1}{7}(-1) + \cdots\right)$$

$$= 4\left(1 - \frac{1}{3} + \frac{1}{5} - \frac{1}{7} + \cdots\right)$$

$$\boxed{\frac{\pi}{4} = 1 - \frac{1}{3} + \frac{1}{5} - \frac{1}{7} + \cdots}$$

8.5.1 How to Choose a Suitable Point

In Example 1 we used point $x = 0$. In Example 2 we used the three points $x = 0$, $x = \pi$ and $x = \frac{\pi}{2}$. How do we know the point to evaluate the Fourier series in order to get cool mathematical results? It is very simple. We will just try the three points:

$$\boxed{\begin{array}{l} x = 0 \\ x = \pi \\ x = \dfrac{\pi}{2} \end{array}}$$

These points are special because sine and cosine functions ($\sin nx$ or $\cos nx$ for integer n) always give whole number values at these points.

Example 3

Consider the function below.

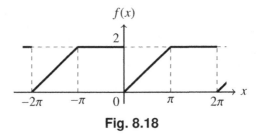

Fig. 8.18

The Fourier series for $f(x)$ is

$$f(x) = \frac{3}{2} - \frac{4}{\pi^2}\left(\cos x + \frac{1}{3^2}\cos 3x + \cdots\right) - \frac{2}{\pi}\left(\sin x + \frac{1}{2}\sin 2x + \cdots\right)$$

Derive π-related formulas from this Fourier series.

Solution   The analytical definition of the function is

$$f(x) = \begin{cases} 2 & -\pi < x < 0 \\ \dfrac{2x}{\pi} & 0 < x < \pi \end{cases}$$

$$f(x + 2\pi) = f(x)$$

We need the analytical definition of the function in order to evaluate its value at $x = \dfrac{\pi}{2}$, or any other point whose value is not obvious from the graph.

Point $x = 0$ From the graph, $x = 0$ is a point of jump discontinuity.

$$\text{top value} = 2$$
$$\text{bottom value} = 0$$
$$\text{average} = \frac{2 + 0}{2}$$
$$= 1$$

Therefore, at $x = 0$, $f(x) = 1$. If we put $x = 0$, $f(x) = 1$ into the Fourier series, we get

$$1 = \frac{3}{2} - \frac{4}{\pi^2}\left(\cos 0 + \frac{1}{3^2}\cos 0 + \cdots\right) - \frac{2}{\pi}\left(\sin 0 + \frac{1}{2}\sin 0 + \cdots\right)$$

$$1 - \frac{3}{2} = -\frac{4}{\pi^2}\left(1 + \frac{1}{3^2}(1) + \cdots\right) - \frac{2}{\pi}\left(0 + 0 + \cdots\right)$$

$$-\frac{1}{2} = -\frac{4}{\pi^2}\left(1 + \frac{1}{3^2}(1) + \cdots\right) - 0$$

$$\boxed{\frac{\pi^2}{8} = 1 + \frac{1}{3^2} + \frac{1}{5^2} + \frac{1}{7^2} + \cdots}$$

Point $x = \pi$ From the graph, at $x = \pi$, $f(x) = 2$. If we put $x = \pi$, $f(x) = 2$ into the Fourier series, we get

$$2 = \frac{3}{2} - \frac{4}{\pi^2}\left(\cos \pi + \frac{1}{3^2}\cos \pi + \cdots\right) - \frac{2}{\pi}\left(\sin \pi + \frac{1}{2}\sin \pi + \cdots\right)$$

$$2 - \frac{3}{2} = -\frac{4}{\pi^2}\left(\cos \pi + \frac{1}{3^2}\cos 3\pi + \cdots\right) - \frac{2}{\pi}\left(0 + 0 + \cdots\right)$$

$$\frac{1}{2} = -\frac{4}{\pi^2}\left(\cos \pi + \frac{1}{3^2}\cos 3\pi + \cdots\right) - 0$$

$$-\frac{\pi^2}{8} = \cos \pi + \frac{1}{3^2}\cos 3\pi + \cdots$$

Recall that

$$\cos n\pi = \begin{cases} -1 & n \text{ is odd} \\ 1 & n \text{ is even} \end{cases}$$

Therefore

$$-\frac{\pi^2}{8} = (-1) + \frac{1}{3^2}(-1) + \frac{1}{5^2}(-1) + \frac{1}{7^2}(-1) + \cdots$$

$$-\frac{\pi^2}{8} = (-1)\left(1 + \frac{1}{3^2} + \frac{1}{5^2} + \frac{1}{7^2} + \cdots\right)$$

$$\boxed{\frac{\pi^2}{8} = 1 + \frac{1}{3^2} + \frac{1}{5^2} + \frac{1}{7^2} + \cdots}$$

This is the same result we got at point $x = 0$.

Point $x = \dfrac{\pi}{2}$ From the analytical definition of $f(x)$,

$$f(x) = \frac{2x}{\pi} \qquad 0 < x < \pi$$

Thus at $x = \dfrac{\pi}{2}$,

$$f(x) = \frac{2(\pi/2)}{\pi}$$

$$= 1$$

If we put $x = \dfrac{\pi}{2}$, $f(x) = 1$ into the Fourier series, we get

$$1 = \frac{3}{2} - \frac{4}{\pi^2}\left(\cos\frac{\pi}{2} + \frac{1}{3^2}\cos 3\left(\frac{\pi}{2}\right) + \cdots\right) - \frac{2}{\pi}\left(\sin\frac{\pi}{2} + \frac{1}{2}\sin 2\left(\frac{\pi}{2}\right) + \cdots\right)$$

$$1 - \frac{3}{2} = -\frac{4}{\pi^2}\left(\cos\frac{\pi}{2} + \frac{1}{3^2}\cos\frac{3\pi}{2} + \cdots\right) - \frac{2}{\pi}\left(\sin\frac{\pi}{2} + \frac{1}{2}\sin\frac{2\pi}{2} + \cdots\right)$$

$$-\frac{1}{2} = -\frac{4}{\pi^2}\left(\cos\frac{\pi}{2} + \frac{1}{3^2}\cos\frac{3\pi}{2} + \cdots\right) - \frac{2}{\pi}\left(\sin\frac{\pi}{2} + \frac{1}{2}\sin\frac{2\pi}{2} + \cdots\right)$$

Recall that

$$\cos\frac{n\pi}{2} = 0 \quad (n \text{ is odd}) \quad \text{and} \quad \sin\frac{n\pi}{2} = \begin{cases} 0 & n \text{ is even} \\ 1 & n = 1, 5, 9, \ldots \\ -1 & n = 3, 7, 11, \ldots \end{cases}$$

Thus

$$-\frac{1}{2} = -\frac{4}{\pi^2}\left(0 + 0 + \cdots\right) - \frac{2}{\pi}\left((1) + \frac{1}{2}(0) + \frac{1}{3}(-1) + \frac{1}{4}(0) + \frac{1}{5}(1) + \cdots\right)$$

$$-\frac{1}{2} = 0 - \frac{2}{\pi}\left((1) + \frac{1}{3}(-1) + \frac{1}{5}(1) + \cdots\right)$$

$$\boxed{\frac{\pi}{4} = 1 - \frac{1}{3} + \frac{1}{5} - \frac{1}{7} + \cdots}$$

Example 4

Consider the function in Fig. 8.19. Assume the period is 2π.

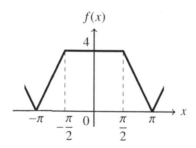

$f(x)$

Fig. 8.19

Given that

$$f(x) = 3 + \frac{16}{\pi^2}\left(\cos x + \frac{1}{3^2}\cos 3x + \cdots\right) - \frac{32}{\pi^2}\left(\frac{1}{2^2}\cos 2x + \frac{1}{6^2}\cos 6x + \cdots\right)$$

derive π-related formulas by putting $x = 0$, $x = \pi$ and $x = \dfrac{\pi}{2}$.

Solution   The analytical definition of the function is

$$
f(x) = \begin{cases} \dfrac{8}{\pi}(x+\pi) & -\pi < x < -\dfrac{\pi}{2} \\[2mm] 4 & -\dfrac{\pi}{2} < x < \dfrac{\pi}{2} \\[2mm] -\dfrac{8}{\pi}(x-\pi) & \dfrac{\pi}{2} < x < \pi \end{cases}
$$

$$f(x + 2\pi) = f(x)$$

Point $x = 0$ From the graph, at $x = 0$, $f(x) = 4$. If we put $x = 0$, $f(x) = 4$ into the Fourier series, we get

$$4 = 3 + \frac{16}{\pi^2}\Big(\cos 0 + \frac{1}{3^2}\cos 0 + \cdots\Big) - \frac{32}{\pi^2}\Big(\frac{1}{2^2}\cos 0 + \frac{1}{6^2}\cos 0 + \cdots\Big)$$

$$4 - 3 = \frac{16}{\pi^2}\Big(1 + \frac{1}{3^2}(1) + \cdots\Big) - \frac{32}{\pi^2}\Big(\frac{1}{2^2}(1) + \frac{1}{6^2}(1) + \cdots\Big)$$

$$1 = \frac{16}{\pi^2}\Big(1 + \frac{1}{3^2} + \frac{1}{5^2} + \cdots\Big) - \frac{32}{\pi^2}\Big(\frac{1}{2^2} + \frac{1}{6^2} + \frac{1}{10^2} + \cdots\Big)$$

Recall that in Example 3, we proved that

$$1 + \frac{1}{3^2} + \frac{1}{5^2} + \frac{1}{7^2} + \cdots = \frac{\pi^2}{8}$$

Thus

$$1 = \frac{16}{\pi^2}\Big(\frac{\pi^2}{8}\Big) - \frac{32}{\pi^2}\Big(\frac{1}{2^2} + \frac{1}{6^2} + \frac{1}{10^2} + \cdots\Big)$$

$$1 = 2 - \frac{32}{\pi^2}\Big(\frac{1}{2^2} + \frac{1}{6^2} + \frac{1}{10^2} + \cdots\Big)$$

$$-1 = -\frac{32}{\pi^2}\Big(\frac{1}{2^2} + \frac{1}{6^2} + \frac{1}{10^2} + \cdots\Big)$$

$$\boxed{\frac{\pi^2}{32} = \frac{1}{2^2} + \frac{1}{6^2} + \frac{1}{10^2} + \cdots}$$

Point $x = \pi$ From the graph, at $x = \pi$, $f(x) = 0$. If we put $x = \pi$, $f(x) = 0$ into the Fourier series, we get

$$0 = 3 + \frac{16}{\pi^2}\left(\cos \pi + \frac{1}{3^2}\cos 3\pi + \cdots\right) - \frac{32}{\pi^2}\left(\frac{1}{2^2}\cos 2\pi + \frac{1}{6^2}\cos 6\pi + \cdots\right)$$

$$-3 = \frac{16}{\pi^2}\left(\cos \pi + \frac{1}{3^2}\cos 3\pi + \cdots\right) - \frac{32}{\pi^2}\left(\frac{1}{2^2}\cos 2\pi + \frac{1}{6^2}\cos 6\pi + \cdots\right)$$

Recall that

$$\cos n\pi = \begin{cases} -1 & n \text{ is odd} \\ 1 & n \text{ is even} \end{cases}$$

Thus

$$-3 = \frac{16}{\pi^2}\left((-1) + \frac{1}{3^2}(-1) + \cdots\right) - \frac{32}{\pi^2}\left(\frac{1}{2^2}(1) + \frac{1}{6^2}(1) + \cdots\right)$$

$$-3 = -\frac{16}{\pi^2}\left(1 + \frac{1}{3^2} + \cdots\right) - \frac{32}{\pi^2}\left(\frac{1}{2^2} + \frac{1}{6^2} + \cdots\right)$$

$$3 = \frac{16}{\pi^2}\left(1 + \frac{1}{3^2} + \cdots\right) + \frac{32}{\pi^2}\left(\frac{1}{2^2} + \frac{1}{6^2} + \cdots\right)$$

multiply through by $\dfrac{\pi^2}{16}$

$$\boxed{\frac{3\pi^2}{16} = \left(1 + \frac{1}{3^2} + \frac{1}{5^2} + \cdots\right) + 2\left(\frac{1}{2^2} + \frac{1}{6^2} + \frac{1}{10^2} + \cdots\right)}$$

Point $x = \dfrac{\pi}{2}$ From the graph, at $x = \dfrac{\pi}{2}$, $f(x) = 4$. If we put $x = \dfrac{\pi}{2}$, $f(x) = 4$ into the Fourier series, we get

$$4 = 3 + \frac{16}{\pi^2}\left(\cos \frac{\pi}{2} + \frac{1}{3^2}\cos 3\left(\frac{\pi}{2}\right) + \cdots\right)$$

$$- \frac{32}{\pi^2}\left(\frac{1}{2^2}\cos 2\left(\frac{\pi}{2}\right) + \frac{1}{6^2}\cos 6\left(\frac{\pi}{2}\right) + \cdots\right)$$

$$4 - 3 = \frac{16}{\pi^2}\left(\cos \frac{\pi}{2} + \frac{1}{3^2}\cos \frac{3\pi}{2} + \cdots\right) - \frac{32}{\pi^2}\left(\frac{1}{2^2}\cos \frac{2\pi}{2} + \frac{1}{6^2}\cos \frac{6\pi}{2} + \cdots\right)$$

$$1 = \frac{16}{\pi^2}\left(\cos \frac{\pi}{2} + \frac{1}{3^2}\cos \frac{3\pi}{2} + \cdots\right) - \frac{32}{\pi^2}\left(\frac{1}{2^2}\cos \frac{2\pi}{2} + \frac{1}{6^2}\cos \frac{6\pi}{2} + \cdots\right)$$

Recall that

$$\cos \frac{n\pi}{2} = \begin{cases} 0 & n \text{ is odd} \\ -1 & n = 2, 6, 10, \ldots \end{cases}$$

Thus

$$1 = \frac{16}{\pi^2}(0 + 0 + \cdots) - \frac{32}{\pi^2}\left(\frac{1}{2^2}(-1) + \frac{1}{6^2}(-1) + \cdots\right)$$

$$1 = 0 + \frac{32}{\pi^2}\left(\frac{1}{2^2} + \frac{1}{6^2} + \cdots\right)$$

$$\boxed{\frac{\pi^2}{32} = \frac{1}{2^2} + \frac{1}{6^2} + \frac{1}{10^2} + \cdots}$$

Exercise 1

For each function below derive a formula for π by putting $x = 0$, $x = \pi$ and $x = \frac{\pi}{2}$ into the Fourier series of the function. All functions have period 2π.

1.

$$f(x) = \begin{cases} -6 & -\pi < x < 0 \\ 6 & 0 < x < \pi \end{cases}$$

$$f(x) = \frac{24}{\pi}\left(\sin x + \frac{1}{3}\sin 3x + \cdots\right)$$

2.

$$f(x) = \begin{cases} 0 & -\pi < x < 0 \\ 6 & 0 < x < \pi \end{cases}$$

$$f(x) = 3 + \frac{24}{\pi}\left(\sin x + \frac{1}{3}\sin 3x + \cdots\right)$$

3.

$$f(x) = x \qquad 0 < x < 2\pi$$

$$f(x) = \pi - 2\left(\sin x + \frac{1}{2}\sin 2x + \cdots\right)$$

4.

$$f(x) = \begin{cases} x & -\frac{\pi}{2} < x < \frac{\pi}{2} \\ \pi - x & \frac{\pi}{2} < x < \frac{3\pi}{2} \end{cases}$$

$$f(x) = \frac{4}{\pi}\left(\sin x - \frac{1}{3^2}\sin 3x + \cdots\right)$$

5.

$$f(x) = \begin{cases} \pi + x & -\pi < x < -\frac{\pi}{2} \\ \frac{\pi}{2} & -\frac{\pi}{2} < x < \frac{\pi}{2} \\ \pi - x & \frac{\pi}{2} < x < \pi \end{cases}$$

$$f(x) = \frac{3\pi}{8} + \frac{2}{\pi}\left(\cos x + \frac{1}{3^2}\cos 3x + \cdots\right)$$
$$- \frac{4}{\pi}\left(\frac{1}{2^2}\cos 2x + \frac{1}{6^2}\cos 6x + \cdots\right)$$

Exercise 2

For each function below, derive a formula for π by putting $x = 0$, $x = \pi$ and $x = \dfrac{\pi}{2}$ into the Fourier series of the function. All functions have period 2π.

1.

$f(x)$

$$f(x) = \sin x + \frac{1}{3}\sin 3x + \frac{1}{5}\sin 5x + \cdots$$

3.

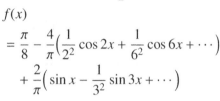

$f(x)$

$$= \frac{\pi}{8} - \frac{4}{\pi}\left(\frac{1}{2^2}\cos 2x + \frac{1}{6^2}\cos 6x + \cdots\right)$$
$$+ \frac{2}{\pi}\left(\sin x - \frac{1}{3^2}\sin 3x + \cdots\right)$$

2.

$f(x)$

$$f(x) = \frac{\pi}{2} - \frac{4}{\pi}\left(\cos x + \frac{1}{3^2}\cos 3x + \cdots\right)$$

8.6   The Basel Problem

The Basel Problem asks for the precise summation of the infinite series:

$$\frac{1}{1^2} + \frac{1}{2^2} + \frac{1}{3^2} + \frac{1}{4^2} + \cdots$$

This is the sum of the reciprocals of the squares of the natural numbers.

The problem was first posed in 1644 and remained open for 90 years, until Euler solved it. Euler's solution brought him fame in the mathematical community. The problem is named after Basel, hometown of Euler.

8.6.1 Solution of the Basel Problem

We can solve the Basel Problem by finding the value of a Fourier series at a point. Consider the function:

$$f(x) = \begin{cases} -x & -\pi < x < 0 \\ x & 0 < x < \pi \end{cases}$$

$$f(x + 2\pi) = f(x)$$

The graph of the function is shown below.

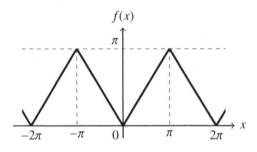

Fig. 8.20

The Fourier series of $f(x)$ is

$$f(x) = \frac{\pi}{2} - \frac{4}{\pi}\left(\cos x + \frac{1}{3^2}\cos 3x + \cdots\right)$$

Point $x = 0$ From the graph, at $x = 0$, $f(x) = 0$. If we put $x = 0$, $f(x) = 0$ into the Fourier series, we get

$$0 = \frac{\pi}{2} - \frac{4}{\pi}\left(\cos 0 + \frac{1}{3^2}\cos 0 + \cdots\right)$$

$$-\frac{\pi}{2} = -\frac{4}{\pi}\left(1 + \frac{1}{3^2}(1) + \cdots\right)$$

$$\boxed{\frac{\pi^2}{8} = 1 + \frac{1}{3^2} + \frac{1}{5^2} + \frac{1}{7^2} + \cdots}$$

Proof Recall that

$$\sum \text{Natural numbers} = \sum \text{Odd numbers} + \sum \text{Even numbers}$$

Similarly

$$\sum \begin{matrix} \text{Natural number} \\ \text{square reciprocals} \end{matrix} = \sum \begin{matrix} \text{Odd number} \\ \text{square reciprocals} \end{matrix} + \sum \begin{matrix} \text{Even number} \\ \text{square reciprocals} \end{matrix}$$

Now we can represent the nth natural number by n and the nth even number by $2n$. Therefore the nth natural number square reciprocal is $\dfrac{1}{n^2}$ and the nth even number square reciprocal is $\dfrac{1}{(2n)^2}$. Therefore we can write the last equation as

$$\sum \frac{1}{n^2} = \sum \begin{matrix} \text{Odd number} \\ \text{square reciprocals} \end{matrix} + \sum \frac{1}{(2n)^2}$$

$$\sum \frac{1}{n^2} - \sum \frac{1}{(2n)^2} = \sum \begin{matrix} \text{Odd number} \\ \text{square reciprocals} \end{matrix}$$

Therefore we have that

$$\sum_{n=1}^{\infty} \frac{1}{n^2} - \sum_{n=1}^{\infty} \frac{1}{(2n)^2} = 1 + \frac{1}{3^2} + \frac{1}{5^2} + \frac{1}{7^2} + \cdots$$

$$\sum_{n=1}^{\infty} \frac{1}{n^2} - \sum_{n=1}^{\infty} \frac{1}{4n^2} = 1 + \frac{1}{3^2} + \frac{1}{5^2} + \frac{1}{7^2} + \cdots$$

$$\sum_{n=1}^{\infty} \left(\frac{1}{n^2} - \frac{1}{4n^2} \right) = 1 + \frac{1}{3^2} + \frac{1}{5^2} + \frac{1}{7^2} + \cdots$$

$$\sum_{n=1}^{\infty} \frac{3}{4n^2} = 1 + \frac{1}{3^2} + \frac{1}{5^2} + \frac{1}{7^2} + \cdots$$

$$\sum_{n=1}^{\infty} \frac{1}{n^2} = \frac{4}{3} \left(1 + \frac{1}{3^2} + \frac{1}{5^2} + \frac{1}{7^2} + \cdots \right)$$

$$\text{recall that } 1 + \frac{1}{3^2} + \cdots = \frac{\pi^2}{8}$$

$$\sum_{n=1}^{\infty} \frac{1}{n^2} = \frac{4}{3}\left(\frac{\pi^2}{8}\right)$$

$$\sum_{n=1}^{\infty} \frac{1}{n^2} = \frac{\pi^2}{6}$$

$$\boxed{1 + \frac{1}{2^2} + \frac{1}{3^2} + \frac{1}{4^2} + \cdots = \frac{\pi^2}{6}}$$

This completes the solution of the Basel Problem.

Author's Note

Congratulations on finishing this book! You have taken a big step in your journey into the amazing world of Fourier series. Thank you so much for taking your time to read this book.

Leave a review on Amazon and let others know how much this book has helped you. This goes a long way to help me as an author. I will really appreciate it. You can be sure I will be there to read your review.

APPENDIX A

Summary of Rules of Int-Vectors

This is a list of all the laws of int-vectors. You can use this section as a reference when solving Fourier series problems. It is advisable that you make a note of these important laws. The results about trigonometric functions, which is used frequently in problems, is listed in the next appendix; also make a note of those.

A.1   General Laws of Int-Vectors

Rule 1 $(a, b) = (a, 0) + (0, b)$

Rule 2 $(a, b) = -(b, a)$

A.2   Int-Vectors for Finding Fourier Coefficient

A.2.1   Rules when solving for a_n

Rule 3a $(0, a) = \dfrac{1}{n\omega} \sin n\omega a$

Rule 4a $x(0, a) = \dfrac{a}{n\omega} \sin n\omega a + \dfrac{1}{n^2\omega^2}(\cos n\omega a - 1)$

A.2.2   Rules when solving for b_n

Rule 3b $(0, a) = \dfrac{1}{n\omega}(1 - \cos n\omega a)$

Rule 4b $x(0, a) = -\dfrac{a}{n\omega} \cos n\omega a + \dfrac{1}{n^2\omega^2} \sin n\omega a$

A.2.3   Rules when solving for a_0

Rule 3c $(0, a) = a$

Rule 4c $x(0, a) = \dfrac{a^2}{2}$

A.3   Secondary Laws of Int-Vectors

A.3.1   Rules when Solving for a_0 and a_n

For real numbers a, b,

Rule 5a $(-a, -b) = -(a, b)$

Rule 6a $x(-a, -b) = x(a, b)$

A.3.2   Rules when Solving for b_n

For real numbers a, b,

Rule 5b $(-a, -b) = (a, b)$

Rule 6b $x(-a, -b) = -x(a, b)$

APPENDIX B

Trigonometric Functions

In this section we establish a proof for these results:

1.

$$\sin n\pi = 0$$

$$\cos n\pi = \begin{cases} -1 & n \text{ is odd} \\ 1 & n \text{ is even} \end{cases}$$

2.

$$\sin \frac{n\pi}{2} = \begin{cases} 0 & n \text{ is even} \\ 1 & n = 1, 5, 9, \ldots \\ -1 & n = 3, 7, 11, \ldots \end{cases} \qquad \cos \frac{n\pi}{2} = \begin{cases} 0 & n \text{ is odd} \\ -1 & n = 2, 6, 10, \ldots \\ 1 & n = 4, 8, 12, \ldots \end{cases}$$

B.1 The Sine Function

Consider the graph of $f(x) = \sin x$.

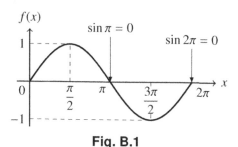

Fig. B.1

The period of $f(x)$ is 2π. Thus

$$f(x + 2\pi) = f(x)$$
$$\sin(x + 2\pi) = \sin x$$
$$\text{put } x = n\pi$$
$$\sin(n\pi + 2\pi) = \sin n\pi$$
$$\sin(n + 2)\pi = \sin n\pi \qquad (B.1)$$

When n is odd   From Fig. B.1 we have that $\sin \pi = 0$, and from Equation (B.1) we can establish that

$$\sin \pi = \sin(1 + 2)\pi = \sin(3 + 2)\pi = \sin(5 + 2)\pi = \cdots$$
$$\therefore \sin \pi = \sin 3\pi = \sin 5\pi = \sin 7\pi = \cdots$$

This is true because consecutive odd numbers differ by 2. Thus

$$0 = \sin \pi = \sin 3\pi = \sin 5\pi = \sin 7\pi = \cdots$$

Therefore

$$\sin n\pi = 0 \quad n \text{ is odd} \tag{B.2}$$

When n is even   From Fig. B.1 we have that $\sin 2\pi = 0$, and from Equation (B.1) we can establish that

$$\sin 2\pi = \sin(2 + 2)\pi = \sin(4 + 2)\pi = \sin(6 + 2)\pi = \cdots$$
$$\therefore \sin 2\pi = \sin 4\pi = \sin 6\pi = \sin 8\pi = \cdots$$

This is true because consecutive even numbers differ by 2. Thus

$$0 = \sin 2\pi = \sin 4\pi = \sin 6\pi = \sin 8\pi = \cdots$$

Therefore

$$\sin n\pi = 0 \quad n \text{ is even} \tag{B.3}$$

Combining results   If we combine (B.2) and (B.3), we get

$$\boxed{\sin n\pi = 0} \quad n = 1, 2, 3, \ldots \tag{B.4}$$

B.1.1   Proof for $\sin \dfrac{n\pi}{2}$

We have that

$$\sin(x + 2\pi) = \sin x$$
$$\text{put } x = \frac{n\pi}{2}$$
$$\sin(\frac{n\pi}{2} + 2\pi) = \sin \frac{n\pi}{2}$$
$$\sin(\frac{n\pi}{2} + \frac{4\pi}{2}) = \sin \frac{n\pi}{2}$$
$$\sin(n + 4)\frac{\pi}{2} = \sin \frac{n\pi}{2} \tag{B.5}$$

When $n = 1, 5, 9, \ldots$ From Fig. B.1 we have that $\sin\frac{\pi}{2} = 1$, and from (B.5) we can establish that

$$\sin\frac{\pi}{2} = \sin(1 + 4)\frac{\pi}{2} = \sin(5 + 4)\frac{\pi}{2} = \sin(9 + 4)\frac{\pi}{2} = \cdots$$

$$\therefore \sin\frac{\pi}{2} = \sin\frac{5\pi}{2} = \sin\frac{9\pi}{2} = \sin\frac{13\pi}{2} = \cdots$$

Thus

$$1 = \sin\frac{\pi}{2} = \sin\frac{5\pi}{2} = \sin\frac{9\pi}{2} = \sin\frac{13\pi}{2} = \cdots$$

Therefore

$$\sin\frac{n\pi}{2} = 1 \quad n = 1, 5, 9, \ldots \tag{B.6}$$

When $n = 2, 6, 10, \ldots$ From Fig. B.1 we have that $\sin\frac{2\pi}{2} = 0$, and from (B.5) we can establish that

$$\sin\frac{2\pi}{2} = \sin(2 + 4)\frac{\pi}{2} = \sin(6 + 4)\frac{\pi}{2} = \sin(10 + 4)\frac{\pi}{2} = \cdots$$

$$\therefore \sin\frac{2\pi}{2} = \sin\frac{6\pi}{2} = \sin\frac{10\pi}{2} = \sin\frac{14\pi}{2} = \cdots$$

Thus

$$0 = \sin\frac{2\pi}{2} = \sin\frac{6\pi}{2} = \sin\frac{10\pi}{2} = \sin\frac{14\pi}{2} = \cdots$$

Therefore

$$\sin\frac{n\pi}{2} = 0 \quad n = 2, 6, 10, \ldots \tag{B.7}$$

Extending the Result Similar to the previous two cases, we have that:

$$\sin\frac{n\pi}{2} = -1 \quad n = 3, 7, 11, \ldots$$

$$\sin\frac{n\pi}{2} = 0 \quad n = 4, 8, 12, \ldots$$

Thus

$$\sin\frac{n\pi}{2} = \begin{cases} 1 & n = 1, 5, 9, \ldots \\ 0 & n = 2, 6, 10, \ldots \\ -1 & n = 3, 7, 11, \ldots \\ 0 & n = 4, 8, 12, \ldots \end{cases}$$

Therefore

$$\sin\frac{n\pi}{2} = \begin{cases} 0 & n \text{ is even} \\ 1 & n = 1, 5, 9, \ldots \\ -1 & n = 3, 7, 11, \ldots \end{cases}$$

B.2   The Cosine Function

Consider the graph of $f(x) = \cos x$.

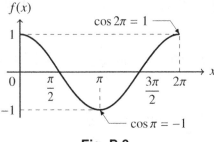

Fig. B.2

The period of $f(x)$ is 2π. Thus

$$f(x + 2\pi) = f(x)$$
$$\cos(x + 2\pi) = \cos x$$
$$\text{put } x = n\pi$$
$$\cos(n\pi + 2\pi) = \cos n\pi$$
$$\cos(n + 2)\pi = \cos n\pi \qquad \text{(B.8)}$$

When n is odd   From Fig. B.2 we have that $\cos\pi = -1$, and from (B.8) we can establish that

$$\cos\pi = \cos(1 + 2)\pi = \cos(3 + 2)\pi = \cos(5 + 2)\pi = \cdots$$
$$\therefore \cos\pi = \cos 3\pi = \cos 5\pi = \cos 7\pi = \cdots$$

This is true because consecutive odd numbers differ by 2. Thus

$$-1 = \cos\pi = \cos 3\pi = \cos 5\pi = \cos 7\pi = \cdots$$

Therefore

$$\cos n\pi = -1 \quad n \text{ is odd} \qquad \text{(B.9)}$$

When n is even From Fig. B.2 we have that $\cos 2\pi = 1$, and from (B.8) we can establish that

$$\cos 2\pi = \cos(2 + 2)\pi = \cos(4 + 2)\pi = \cos(6 + 2)\pi = \cdots$$
$$\therefore \cos 2\pi = \cos 4\pi = \cos 6\pi = \cos 8\pi = \cdots$$

This is true because consecutive even numbers differ by 2. Thus

$$1 = \cos 2\pi = \cos 4\pi = \cos 6\pi = \cos 8\pi = \cdots$$

Therefore

$$\cos n\pi = 1 \quad n \text{ is even} \tag{B.10}$$

Combining results If we combine (B.9) and (B.10) then we get

$$\cos n\pi = \begin{cases} -1 & n \text{ is odd} \\ 1 & n \text{ is even} \end{cases} \tag{B.11}$$

B.2.1 Proof for $\cos \dfrac{n\pi}{2}$

We have that

$$\cos(x + 2\pi) = \cos x$$
$$\text{put } x = \frac{n\pi}{2}$$
$$\cos(\frac{n\pi}{2} + 2\pi) = \cos \frac{n\pi}{2}$$
$$\cos(\frac{n\pi}{2} + \frac{4\pi}{2}) = \cos \frac{n\pi}{2}$$
$$\cos(n + 4)\frac{\pi}{2} = \cos \frac{n\pi}{2} \tag{B.12}$$

When $n = 1, 5, 9, \ldots$ From Fig. B.2 we have that $\cos \dfrac{\pi}{2} = 0$, and from (B.12) we can establish that

$$\cos \frac{\pi}{2} = \cos(1 + 4)\frac{\pi}{2} = \cos(5 + 4)\frac{\pi}{2} = \cos(9 + 4)\frac{\pi}{2} = \cdots$$

$$\therefore \cos \frac{\pi}{2} = \cos \frac{5\pi}{2} = \cos \frac{9\pi}{2} = \cos \frac{13\pi}{2} = \cdots$$

Thus

$$0 = \cos \frac{\pi}{2} = \cos \frac{5\pi}{2} = \cos \frac{9\pi}{2} = \cos \frac{13\pi}{2} = \cdots$$

Therefore

$$\cos \frac{n\pi}{2} = 0 \quad n = 1, 5, 9, \ldots \tag{B.13}$$

When $n = 2, 6, 10, \ldots$ From Fig. B.2 we have that $\cos \dfrac{2\pi}{2} = -1$, and from (B.12) we can establish that

$$\cos \frac{2\pi}{2} = \cos(2 + 4)\frac{\pi}{2} = \cos(6 + 4)\frac{\pi}{2} = \cos(10 + 4)\frac{\pi}{2} = \cdots$$

$$\therefore \cos \frac{2\pi}{2} = \cos \frac{6\pi}{2} = \cos \frac{10\pi}{2} = \cos \frac{14\pi}{2} = \cdots$$

Thus

$$-1 = \cos \frac{2\pi}{2} = \cos \frac{6\pi}{2} = \cos \frac{10\pi}{2} = \cos \frac{14\pi}{2} = \cdots$$

Therefore

$$\cos \frac{n\pi}{2} = -1 \quad n = 2, 6, 10, \ldots \tag{B.14}$$

Extending the Result   Similar to the previous two cases, we have that:

$$\cos \frac{n\pi}{2} = 0 \quad n = 3, 7, 11, \ldots$$

$$\cos \frac{n\pi}{2} = 1 \quad n = 4, 8, 12, \ldots$$

Thus

$$\cos \frac{n\pi}{2} = \begin{cases} 0 & n = 1, 5, 9, \ldots \\ -1 & n = 2, 6, 10, \ldots \\ 0 & n = 3, 7, 11, \ldots \\ 1 & n = 4, 8, 12, \ldots \end{cases}$$

Therefore

$$\cos \frac{n\pi}{2} = \begin{cases} 0 & n \text{ is odd} \\ -1 & n = 2, 6, 10, \ldots \\ 1 & n = 4, 8, 12, \ldots \end{cases}$$

APPENDIX C

The Math Behind Int-Vectors

You must be curious about why int-vectors can be used to solve Fourier series. This section is dedicated to explaining the mathematical background of int-vectors. We establish int-vectors as the definite integral of a function.

C.1   Properties of the Definite Integral

P1. $\displaystyle \int_a^b f(x)\,dx = -\int_b^a f(x)\,dx$

P2. $\displaystyle \int_a^b f(x)\,dx = \int_a^0 f(x)\,dx + \int_0^b f(x)\,dx$

P3. If $f(x)$ is even then $\displaystyle \int_{-a}^{-b} f(x)\,dx = -\int_a^b f(x)\,dx$

P4. If $f(x)$ is odd then $\displaystyle \int_{-a}^{-b} f(x)\,dx = \int_a^b f(x)\,dx$

C.2   Int-Vectors

Let $f(x)$ be an arbitrary function that can be represented as a Fourier series. Then we define the int-vectors (a, b) and $x(a, b)$ thus:

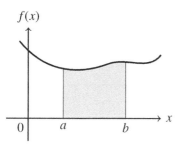

Fig. C.1

V1. $\displaystyle (a, b) = \int_a^b f(x)\,dx$

V2. $\displaystyle x(a, b) = \int_a^b x \cdot f(x)\,dx$

C.3 General Laws of Int-Vectors

Here we shall derive the general laws of int-vectors using the first and second properties (**P1** and **P2**) of definite integrals.

Rule 1   From **P2** of definite integrals

$$\int_a^b f(x)\,dx = \int_a^0 f(x)\,dx + \int_0^b f(x)\,dx$$

Therefore

$$\boxed{(a, b) = (a, 0) + (0, b)}$$

Rule 2   From **P1** of definite integrals

$$\int_a^b f(x)\,dx = -\int_b^a f(x)\,dx$$

Therefore

$$\boxed{(a, b) = -(b, a)}$$

C.4 Int-Vectors for Finding Fourier Coefficients

In this section we derive the rules for finding Fourier coefficients using only the definitions of int-vectors (**V1** and **V2**)

C.4.1 Solving for a_n

Here, $f(x) = \cos n\omega x$.

Rule 3a   From **V1**,

$$(0, a) = \int_0^a f(x)\, dx$$

put $f(x) = \cos n\omega x$

$$= \int_0^a \cos n\omega x\, dx$$

$$\boxed{(0, a) = \frac{1}{n\omega} \sin n\omega a}$$

Rule 4a   From **V2**,

$$x(0, a) = \int_0^a x \cdot f(x)\, dx$$

put $f(x) = \cos n\omega x$

$$= \int_0^a x \cos n\omega x\, dx$$

$$\boxed{x(0, a) = \frac{a}{n\omega} \sin n\omega a + \frac{1}{n^2\omega^2}(\cos n\omega a - 1)}$$

C.4.2   Solving for b_n

Here, $f(x) = \sin n\omega x$.

Rule 3b   From **V1**,

$$(0, a) = \int_0^a f(x)\, dx$$

put $f(x) = \sin n\omega x$

$$= \int_0^a \sin n\omega x\, dx$$

$$\boxed{(0, a) = \frac{1}{n\omega}(1 - \cos n\omega a)}$$

Rule 4b   From **V2**,

$$x(0, a) = \int_0^a x \cdot f(x) \, dx$$

$$\text{put } f(x) = \sin n\omega x$$

$$= \int_0^a x \sin n\omega x \, dx$$

$$x(0, a) = -\frac{a}{n\omega} \cos n\omega a + \frac{1}{n^2\omega^2} \sin n\omega a$$

C.4.3   Solving for a_0

Here, $f(x) = 1$.

Rule 3c   From **V1**,

$$(0, a) = \int_0^a f(x) \, dx$$

$$\text{put } f(x) = 1$$

$$= \int_0^a (1) \, dx$$

$$(0, a) = a$$

Rule 4c   From **V2**,

$$x(0, a) = \int_0^a x \cdot f(x) \, dx$$

$$\text{put } f(x) = 1$$

$$= \int_0^a x \, dx$$

$$x(0, a) = \frac{a^2}{2}$$

C.5   Secondary Laws of Int-Vectors

In this section we derive the secondary laws of int-vectors using the definite integral. The Rules 3c and 4c will not be mentioned here because they have been derived in the previous section.

C.5.1   Solving for a_0 and a_n

Here, $f(x) = 1$ or $f(x) = \cos n\omega x$, due to a_0 and a_n respectively. It is clear that $f(x)$ is even in both cases. Thus $x \cdot f(x)$, being a multiple of an odd and an even function, is odd in both cases.

Rule 5a   From **V1**,

$$(-a, -b) = \int_{-a}^{-b} f(x)\,dx$$

since $f(x)$ is even, apply **P3**

$$= -\int_{a}^{b} f(x)\,dx$$

$$= -(a, b)$$

Therefore

$$\boxed{(-a, -b) = -(a, b)}$$

Rule 6a   From **V2**,

$$x(-a, -b) = \int_{-a}^{-b} x \cdot f(x)\,dx$$

since $x \cdot f(x)$ is odd, apply **P4**

$$= \int_{a}^{b} x \cdot f(x)\,dx$$

$$= x(a, b)$$

Therefore

$$\boxed{x(-a, -b) = x(a, b)}$$

C.5.2 Solving for b_n

Here, $f(x) = \sin n\omega x$. It is clear that $f(x)$ is odd. Thus $x \cdot f(x)$, being a multiple of two odd functions, is even.

Rule 5b   From **V1**,

$$(-a, -b) = \int_{-a}^{-b} f(x)\,dx$$

since $f(x)$ is odd, apply **P4**

$$= \int_{a}^{b} f(x)\,dx$$

$$= (a, b)$$

Therefore

$$\boxed{(-a, -b) = (a, b)}$$

Rule 6b   From **V2**,

$$x(-a, -b) = \int_{-a}^{-b} x \cdot f(x)\,dx$$

since $x \cdot f(x)$ is even, apply **P3**

$$= -\int_{a}^{b} x \cdot f(x)\,dx$$

$$= -x(a, b)$$

Therefore

$$\boxed{x(-a, -b) = -x(a, b)}$$

Full Worked Solutions to Exercises

Chapter 1

Exercise 1

1. $\sin 2n\pi$

Recall that $\sin n\pi = 0$ for integer n. Since $2n$ is an integer, then

$$\boxed{\sin 2n\pi = 0}$$

2. $\sin 11n\pi$

Recall that $\sin n\pi = 0$ for integer n. Since $11n$ is an integer, then

$$\boxed{\sin 11n\pi = 0}$$

3. $\cos 2n\pi$

Recall that

$$\cos n\pi = \begin{cases} -1 & n \text{ is odd} \\ 1 & \boxed{n \text{ is even}} \end{cases}$$

Since $2n$ is an even integer, then

$$\boxed{\cos 2n\pi = 1}$$

4. $\cos 12n\pi$

Recall that

$$\cos n\pi = \begin{cases} -1 & n \text{ is odd} \\ 1 & \boxed{n \text{ is even}} \end{cases}$$

Since $12n$ is an even integer, then

$$\boxed{\cos 12n\pi = 1}$$

5. $\sin n\pi + \cos 6n\pi$

Recall that

$$\sin n\pi = 0$$

and

$$\cos n\pi = \begin{cases} -1 & n \text{ is odd} \\ 1 & n \text{ is even} \end{cases}$$

Since $6n$ is an even integer, then $\cos 6n\pi = 1$

Therefore $\sin n\pi + \cos 6n\pi = 0 + 1 = 1$

$$\boxed{\sin n\pi + \cos 6n\pi = 1}$$

Exercise 2

1. $1 - \cos n\pi$

Recall that

$$\cos n\pi = \begin{cases} -1 & n \text{ is odd} \\ 1 & n \text{ is even} \end{cases}$$

When n is odd

$$1 - \cos n\pi = 1 - (-1)$$
$$= 1 + 1$$
$$= 2$$

When n is even

$$1 - \cos n\pi = 1 - 1$$
$$= 0$$

Therefore

$$\boxed{1 - \cos n\pi = \begin{cases} 2 & n \text{ is odd} \\ 0 & n \text{ is even} \end{cases}}$$

2. $\dfrac{1}{n^2}\sin n\pi$

Put $\sin n\pi = 0$

$$\therefore \quad \frac{1}{n^2}\sin n\pi = \frac{1}{n^2}(0)$$

$$\boxed{\frac{1}{n^2}\sin n\pi = 0}$$

3. $\dfrac{1}{n\pi}(\cos n\pi - 1)$

Recall that

$$\cos n\pi = \begin{cases} -1 & n \text{ is odd} \\ 1 & n \text{ is even} \end{cases}$$

When n is odd

$$\frac{1}{n\pi}(\cos n\pi - 1) = \frac{1}{n\pi}(-1 - 1)$$
$$= \frac{1}{n\pi}(-2)$$
$$= -\frac{2}{n\pi}$$

When n is even

$$\frac{1}{n\pi}(\cos n\pi - 1) = \frac{1}{n\pi}(1 - 1)$$
$$= \frac{1}{n\pi}(0)$$
$$= 0$$

Therefore

$$\boxed{\frac{1}{n\pi}(\cos n\pi - 1) = \begin{cases} -\dfrac{2}{n\pi} & n \text{ is odd} \\ 0 & n \text{ is even} \end{cases}}$$

4. $\cos n\pi - \cos\dfrac{n\pi}{2}$

Recall that

$$\cos n\pi = \begin{cases} -1 & n \text{ is odd} \\ 1 & n \text{ is even} \end{cases}$$

and

$$\cos\frac{n\pi}{2} = \begin{cases} 0 & n \text{ is odd} \\ -1 & n = 2, 6, 10, \ldots \\ 1 & n = 4, 8, 12, \ldots \end{cases}$$

When n is odd

$$\cos n\pi - \cos\frac{n\pi}{2} = -1 - 0$$
$$= -1$$

When $n = 2, 6, 10, \ldots$

$$\cos n\pi - \cos\frac{n\pi}{2} = 1 - (-1)$$
$$= 1 + 1$$
$$= 2$$

When $n = 4, 8, 12, \ldots$

$$\cos n\pi - \cos\frac{n\pi}{2} = 1 - 1$$
$$= 0$$

Therefore

$$\boxed{\cos n\pi - \cos\frac{n\pi}{2} = \begin{cases} -1 & n \text{ is odd} \\ 2 & n = 2, 6, 10, \ldots \\ 0 & n = 4, 8, 12, \ldots \end{cases}}$$

5. $\dfrac{1}{n}\left(\cos\dfrac{n\pi}{2} - \cos n\pi\right)$

Recall that

$$\cos n\pi = \begin{cases} -1 & n \text{ is odd} \\ 1 & n \text{ is even} \end{cases}$$

and

$$\cos\frac{n\pi}{2} = \begin{cases} 0 & n \text{ is odd} \\ -1 & n = 2, 6, 10, \ldots \\ 1 & n = 4, 8, 12, \ldots \end{cases}$$

When n is odd

$$\frac{1}{n}\left(\cos\frac{n\pi}{2} - \cos n\pi\right) = \frac{1}{n}[0 - (-1)]$$

$$= \frac{1}{n}(1)$$

$$= \frac{1}{n}$$

When $n = 2, 6, 10, \ldots$

$$\frac{1}{n}\left(\cos\frac{n\pi}{2} - \cos n\pi\right) = \frac{1}{n}(-1 - 1)$$

$$= \frac{1}{n}(-2)$$

$$= -\frac{2}{n}$$

When $n = 4, 8, 12, \ldots$

$$\frac{1}{n}\left(\cos\frac{n\pi}{2} - \cos n\pi\right) = \frac{1}{n}(1 - 1)$$

$$= \frac{1}{n}(0)$$

$$= 0$$

Therefore

$$\frac{1}{n}\left(\cos\frac{n\pi}{2} - \cos n\pi\right)$$

$$= \begin{cases} \dfrac{1}{n} & n \text{ is odd} \\ -\dfrac{2}{n} & n = 2, 6, 10, \ldots \\ 0 & n = 4, 8, 12, \ldots \end{cases}$$

6. $-\dfrac{2\pi}{n}\cos 2n\pi + \dfrac{1}{n^2}\sin 2n\pi$

Recall that $\cos 2n\pi = 1$ because $2n$ is an even integer, and $\sin 2n\pi = 0$ because $2n$ is an integer.

Thus

$$-\frac{2\pi}{n}\cos 2n\pi + \frac{1}{n^2}\sin 2n\pi$$

$$= -\frac{2\pi}{n}(1) + \frac{1}{n^2}(0)$$

$$= -\frac{2\pi}{n}$$

Therefore

$$-\frac{2\pi}{n}\cos 2n\pi + \frac{1}{n^2}\sin 2n\pi = -\frac{2\pi}{n}$$

7. $\dfrac{8}{n\pi}\sin\dfrac{n\pi}{2}$

Recall that

$$\sin\frac{n\pi}{2} = \begin{cases} 0 & n \text{ is even} \\ 1 & n = 1, 5, 9, \ldots \\ -1 & n = 3, 7, 11, \ldots \end{cases}$$

Thus

$$\frac{8}{n\pi}\sin\frac{n\pi}{2}$$

$$= \frac{8}{n\pi} \times \begin{cases} 0 & n \text{ is even} \\ 1 & n = 1, 5, 9, \ldots \\ -1 & n = 3, 7, 11, \ldots \end{cases}$$

$$= \begin{cases} \dfrac{8}{n\pi}(0) & n \text{ is even} \\ \dfrac{8}{n\pi}(1) & n = 1, 5, 9, \ldots \\ \dfrac{8}{n\pi}(-1) & n = 3, 7, 11, \ldots \end{cases}$$

Therefore

$$\frac{8}{n\pi}\sin\frac{n\pi}{2} = \begin{cases} 0 & n \text{ is even} \\ \dfrac{8}{n\pi} & n = 1, 5, 9, \ldots \\ -\dfrac{8}{n\pi} & n = 3, 7, 11, \ldots \end{cases}$$

8. $-\dfrac{4}{n\pi}\cos\dfrac{n\pi}{2}$

Recall that

$$\cos\frac{n\pi}{2} = \begin{cases} 0 & n \text{ is odd} \\ -1 & n = 2, 6, 10, \ldots \\ 1 & n = 4, 8, 12, \ldots \end{cases}$$

Thus

$$-\frac{4}{n\pi}\cos\frac{n\pi}{2}$$

$$= -\frac{4}{n\pi} \times \begin{cases} 0 & n \text{ is odd} \\ -1 & n = 2, 6, 10, \ldots \\ 1 & n = 4, 8, 12, \ldots \end{cases}$$

$$= \begin{cases} -\dfrac{4}{n\pi}(0) & n \text{ is odd} \\ -\dfrac{4}{n\pi}(-1) & n = 2, 6, 10, \ldots \\ -\dfrac{4}{n\pi}(1) & n = 4, 8, 12, \ldots \end{cases}$$

Therefore

$$\boxed{\begin{aligned} &-\frac{4}{n\pi}\cos\frac{n\pi}{2} \\ &= \begin{cases} 0 & n \text{ is odd} \\ \dfrac{4}{n\pi} & n = 2, 6, 10, \ldots \\ -\dfrac{4}{n\pi} & n = 4, 8, 12, \ldots \end{cases} \end{aligned}}$$

9. $-\dfrac{2}{n\pi}\sin\dfrac{n\pi}{2}$

Recall that

$$\sin\frac{n\pi}{2} = \begin{cases} 0 & n \text{ is even} \\ 1 & n = 1, 5, 9, \ldots \\ -1 & n = 3, 7, 11, \ldots \end{cases}$$

Thus

$$-\frac{2}{n\pi}\sin\frac{n\pi}{2}$$

$$= -\frac{2}{n\pi} \times \begin{cases} 0 & n \text{ is even} \\ 1 & n = 1, 5, 9, \ldots \\ -1 & n = 3, 7, 11, \ldots \end{cases}$$

$$= \begin{cases} -\dfrac{2}{n\pi}(0) & n \text{ is even} \\ -\dfrac{2}{n\pi}(1) & n = 1, 5, 9, \ldots \\ -\dfrac{2}{n\pi}(-1) & n = 3, 7, 11, \ldots \end{cases}$$

Therefore

$$\boxed{-\frac{2}{n\pi}\sin\frac{n\pi}{2} = \begin{cases} 0 & n \text{ is even} \\ -\dfrac{2}{n\pi} & n = 1, 5, 9, \ldots \\ \dfrac{2}{n\pi} & n = 3, 7, 11, \ldots \end{cases}}$$

10. $-\dfrac{4}{n^2\pi^2}(\cos n\pi - 1)$

Recall that

$$\cos n\pi = \begin{cases} -1 & n \text{ is odd} \\ 1 & n \text{ is even} \end{cases}$$

When n is odd

$$-\frac{4}{n^2\pi^2}(\cos n\pi - 1)$$

$$= -\frac{4}{n^2\pi^2}(-1 - 1)$$

$$= -\frac{4}{n^2\pi^2}(-2)$$

$$= \frac{8}{n^2\pi^2}$$

When n is even

$$-\frac{4}{n^2\pi^2}(\cos n\pi - 1)$$

$$= -\frac{4}{n^2\pi^2}(1 - 1)$$

$$= -\frac{4}{n^2\pi^2}(0)$$

$$= 0$$

Therefore

$$\boxed{-\frac{4}{n^2\pi^2}(\cos n\pi - 1) = \begin{cases} \dfrac{8}{n^2\pi^2} & n \text{ is odd} \\ 0 & n \text{ is even} \end{cases}}$$

11. $\dfrac{2}{n\pi}\sin\dfrac{n\pi}{2} + \dfrac{4}{n^2\pi^2}(\cos n\pi - 1)$

Recall that

$$\sin \frac{n\pi}{2} = \begin{cases} 0 & n \text{ is even} \\ 1 & n = 1, 5, 9, \dots \\ -1 & n = 3, 7, 11, \dots \end{cases}$$

and

$$\cos n\pi = \begin{cases} -1 & n \text{ is odd} \\ 1 & n \text{ is even} \end{cases}$$

When n is even

$$\frac{2}{n\pi} \sin \frac{n\pi}{2} + \frac{4}{n^2\pi^2}(\cos n\pi - 1)$$

$$= \frac{2}{n\pi}(0) + \frac{4}{n^2\pi^2}(1 - 1)$$

$$= 0 + \frac{4}{n^2\pi^2}(0)$$

$$= 0$$

When $n = 1, 5, 9, \dots$

$$\frac{2}{n\pi} \sin \frac{n\pi}{2} + \frac{4}{n^2\pi^2}(\cos n\pi - 1)$$

$$= \frac{2}{n\pi}(1) + \frac{4}{n^2\pi^2}(-1 - 1)$$

$$= \frac{2}{n\pi} + \frac{4}{n^2\pi^2}(-2)$$

$$= \frac{2}{n\pi} - \frac{8}{n^2\pi^2}$$

When $n = 3, 7, 11, \dots$

$$\frac{2}{n\pi} \sin \frac{n\pi}{2} + \frac{4}{n^2\pi^2}(\cos n\pi - 1)$$

$$= \frac{2}{n\pi}(-1) + \frac{4}{n^2\pi^2}(-1 - 1)$$

$$= -\frac{2}{n\pi} + \frac{4}{n^2\pi^2}(-2)$$

$$= -\frac{2}{n\pi} - \frac{8}{n^2\pi^2}$$

Therefore

$$\boxed{\begin{aligned} &\frac{2}{n\pi} \sin \frac{n\pi}{2} + \frac{4}{n^2\pi^2}(\cos n\pi - 1) \\ &= \begin{cases} 0 & n \text{ is even} \\ \dfrac{2}{n\pi} - \dfrac{8}{n^2\pi^2} & n = 1, 5, 9, \dots \\ -\dfrac{2}{n\pi} - \dfrac{8}{n^2\pi^2} & n = 3, 7, 11, \dots \end{cases} \end{aligned}}$$

12. $\dfrac{2}{n^2\pi}\left(2 \cos \dfrac{n\pi}{2} - \cos n\pi - 1\right)$

Recall that

$$\cos \frac{n\pi}{2} = \begin{cases} 0 & n \text{ is odd} \\ -1 & n = 2, 6, 10, \dots \\ 1 & n = 4, 8, 12, \dots \end{cases}$$

and

$$\cos n\pi = \begin{cases} -1 & n \text{ is odd} \\ 1 & n \text{ is even} \end{cases}$$

When n is odd

$$\frac{2}{n^2\pi}\left(2 \cos \frac{n\pi}{2} - \cos n\pi - 1\right)$$

$$= \frac{2}{n^2\pi}[2(0) - (-1) - 1]$$

$$= \frac{2}{n^2\pi}(0 + 1 - 1)$$

$$= \frac{2}{n^2\pi}(0)$$

$$= 0$$

When $n = 2, 6, 10, \dots$

$$\frac{2}{n^2\pi}\left(2 \cos \frac{n\pi}{2} - \cos n\pi - 1\right)$$

$$= \frac{2}{n^2\pi}[2(-1) - 1 - 1]$$

$$= \frac{2}{n^2\pi}(-2 - 2)$$

$$= \frac{2}{n^2\pi}(-4)$$

$$= -\frac{8}{n^2\pi}$$

When $n = 4, 8, 12, \ldots$

$$\frac{2}{n^2\pi}\left(2\cos\frac{n\pi}{2} - \cos n\pi - 1\right)$$

$$= \frac{2}{n^2\pi}[2(1) - 1 - 1]$$

$$= \frac{2}{n^2\pi}(2 - 2)$$

$$= \frac{2}{n^2\pi}(0)$$

$$= 0$$

Therefore

$$\frac{2}{n^2\pi}\left(2\cos\frac{n\pi}{2} - \cos n\pi - 1\right)$$

$$= \begin{cases} 0 & n \text{ is even} \\ -\dfrac{8}{n^2\pi} & n = 2, 6, 10, \ldots \\ 0 & n = 4, 8, 12, \ldots \end{cases}$$

13. $\dfrac{1}{n}(1 - \cos n\pi) - \dfrac{\pi}{n}\cos n\pi + \dfrac{1}{n^2}\sin n\pi$

Recall that

$$\sin n\pi = 0$$

and

$$\cos n\pi = \begin{cases} -1 & n \text{ is odd} \\ 1 & n \text{ is even} \end{cases}$$

Thus

$$\frac{1}{n}(1 - \cos n\pi) - \frac{\pi}{n}\cos n\pi + \frac{1}{n^2}\sin n\pi$$

$$= \frac{1}{n}(1 - \cos n\pi) - \frac{\pi}{n}\cos n\pi + 0$$

$$= \frac{1}{n}(1 - \cos n\pi) - \frac{\pi}{n}\cos n\pi$$

When n **is odd**

$$\frac{1}{n}(1 - \cos n\pi) - \frac{\pi}{n}\cos n\pi$$

$$= \frac{1}{n}[1 - (-1)] - \frac{\pi}{n}(-1)$$

$$= \frac{1}{n}(1 + 1) + \frac{\pi}{n}$$

$$= \frac{1}{n}(2) + \frac{\pi}{n}$$

$$= \frac{2}{n} + \frac{\pi}{n}$$

When n **is even**

$$\frac{1}{n}(1 - \cos n\pi) - \frac{\pi}{n}\cos n\pi$$

$$= \frac{1}{n}(1 - 1) - \frac{\pi}{n}(1)$$

$$= \frac{1}{n}(0) - \frac{\pi}{n}$$

$$= 0 - \frac{\pi}{n}$$

$$= -\frac{\pi}{n}$$

Therefore

$$\frac{1}{n}(1 - \cos n\pi) - \frac{\pi}{n}\cos n\pi + \frac{1}{n^2}\sin n\pi$$

$$= \begin{cases} \dfrac{2}{n} + \dfrac{\pi}{n} & n \text{ is odd} \\ -\dfrac{\pi}{n} & n \text{ is even} \end{cases}$$

Exercise 3

1. If $a_n = \dfrac{8}{n\pi}\sin\dfrac{n\pi}{2}$, evaluate the sum

$$S = \sum_{n=1}^{\infty} a_n \cos nx$$

Recall that

$$\sin\frac{n\pi}{2} = \begin{cases} 0 & n \text{ is even} \\ 1 & n = 1, 5, 9, \ldots \\ -1 & n = 3, 7, 11, \ldots \end{cases}$$

Thus

$$a_n = \begin{cases} 0 & n \text{ is even} \\ \dfrac{8}{n\pi} & n = 1, 5, 9, \ldots \\ -\dfrac{8}{n\pi} & n = 3, 7, 11, \ldots \end{cases}$$

$$S = \sum_{n=1}^{\infty} a_n \cos nx$$

$$= a_n(\cos x + \cos 2x + \cos 3x + \cdots)$$

$$= a_n(\cos 2x + \cos 4x + \cdots)$$
$$+ a_n(\cos x + \cos 5x + \cdots)$$
$$+ a_n(\cos 3x + \cos 7x + \cdots)$$

$$= 0(\cos 2x + \cos 4x + \cdots)$$
$$+ \frac{8}{n\pi}(\cos x + \cos 5x + \cdots)$$
$$- \frac{8}{n\pi}(\cos 3x + \cos 7x + \cdots)$$

$$= 0 + \frac{8}{n\pi}(\cos x + \cos 5x + \cdots)$$
$$- \frac{8}{n\pi}(\cos 3x + \cos 7x + \cdots)$$

$$\boxed{S = \frac{8}{\pi}(\cos x + \frac{1}{5}\cos 5x + \cdots) \\ - \frac{8}{\pi}(\frac{1}{3}\cos 3x + \frac{1}{7}\cos 7x + \cdots)}$$

2. If $b_n = -\dfrac{4}{n^2\pi^2}(\cos n\pi - 1)$, evaluate the sum $S = \sum\limits_{n=1}^{\infty} b_n \sin nx$

Recall that

$$\cos n\pi = \begin{cases} -1 & n \text{ is odd} \\ 1 & n \text{ is even} \end{cases}$$

Thus

$$b_n = \begin{cases} \dfrac{8}{n^2\pi^2} & n \text{ is odd} \\ 0 & n \text{ is even} \end{cases}$$

$$S = \sum_{n=1}^{\infty} b_n \sin nx$$

$$= b_n(\sin x + \sin 2x + \sin 3x + \cdots)$$
$$= b_n(\sin x + \sin 3x + \cdots)$$
$$+ b_n(\sin 2x + \sin 4x + \cdots)$$
$$= \frac{8}{n^2\pi^2}(\sin x + \sin 3x + \cdots)$$
$$+ 0(\sin 2x + \sin 4x + \cdots)$$

$$\boxed{S = \frac{8}{n^2\pi^2}(\sin x + \sin 3x + \cdots)}$$

3. If $a_n = \dfrac{1}{n\pi}(\cos n\pi - 1)$ and $b_n = -\dfrac{4}{n\pi}\cos\dfrac{n\pi}{2}$, evaluate the sum

$$S = \sum_{n=1}^{\infty} a_n \cos nx + \sum_{n=1}^{\infty} b_n \sin nx$$

Recall that

$$\cos n\pi = \begin{cases} -1 & n \text{ is odd} \\ 1 & n \text{ is even} \end{cases}$$

and

$$\cos\frac{n\pi}{2} = \begin{cases} 0 & n \text{ is odd} \\ -1 & n = 2, 6, 10, \ldots \\ 1 & n = 4, 8, 12, \ldots \end{cases}$$

Thus

$$a_n = \begin{cases} -\dfrac{2}{n\pi} & n \text{ is odd} \\ 0 & n \text{ is even} \end{cases}$$

and

$$b_n = \begin{cases} 0 & n \text{ is odd} \\ \dfrac{4}{n\pi} & n = 2, 6, 10, \ldots \\ -\dfrac{4}{n\pi} & n = 4, 8, 12, \ldots \end{cases}$$

$$S = \sum_{n=1}^{\infty} a_n \cos nx + \sum_{n=1}^{\infty} b_n \sin nx$$
$$= a_n(\cos x + \cos 2x + \cdots)$$
$$+ b_n(\sin x + \cos 2x + \cdots)$$
$$= a_n(\cos x + \cos 3x + \cdots)$$
$$+ a_n(\cos 2x + \cos 4x + \cdots)$$
$$+ b_n(\sin x + \cos 3x + \cdots)$$
$$+ b_n(\sin 2x + \cos 6x + \cdots)$$
$$+ b_n(\sin 4x + \cos 8x + \cdots)$$

$$= -\frac{2}{n\pi}(\cos x + \cos 3x + \cdots)$$

$$+ 0(\cos 2x + \cos 4x + \cdots)$$

$$+ 0(\sin x + \cos 3x + \cdots)$$

$$+ \frac{4}{n\pi}(\sin 2x + \cos 6x + \cdots)$$

$$- \frac{4}{n\pi}(\sin 4x + \cos 8x + \cdots)$$

$$= -\frac{2}{\pi}(\cos x + \frac{1}{3}\cos 3x + \cdots)$$

$$+ \frac{4}{\pi}(\frac{1}{2}\sin 2x + \frac{1}{6}\cos 6x + \cdots)$$

$$- \frac{4}{\pi}(\frac{1}{4}\sin 4x + \frac{1}{8}\cos 8x + \cdots)$$

$$\boxed{\begin{aligned} S = -\frac{2}{\pi}(\cos x + \frac{1}{3}\cos 3x + \cdots) \\ + \frac{4}{\pi}(\frac{1}{2}\sin 2x - \frac{1}{4}\sin 4x + \cdots) \end{aligned}}$$

Chapter 2

1.

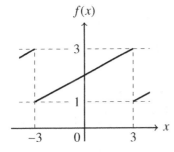

$f(x)$

The graph shows that the period of the function is 6 units. Therefore

$$f(x + 6) = f(x) \qquad \text{(C.1)}$$

Between $x = -3$ and $x = 3$, we have a single straight line which passes through two points:

$$(-3, 1) \quad \text{and} \quad (3, 3)$$

Thus

$$y = \left(\frac{y_1 - y_2}{x_1 - x_2}\right)(x - x_1) + y_1$$

$$= \left(\frac{1 - 3}{-3 - 3}\right)[x - (-3)] + 1$$

$$= \left(\frac{-2}{-6}\right)(x + 3) + 1$$

$$= \left(\frac{1}{3}\right)(x + 3) + 1$$

$$= \frac{1}{3}x + 1 + 1$$

$$= \frac{x}{3} + 2$$

Thus

$$f(x) = \frac{x}{3} + 2 \quad (-3 < x < 3) \qquad \text{(C.2)}$$

From (C.1) and (C.2) we can write the complete analytical definition of $f(x)$:

$$\boxed{\begin{aligned} f(x) = \frac{x}{3} + 2 \quad (-3 < x < 3) \\ f(x + 6) = f(x) \end{aligned}}$$

2.

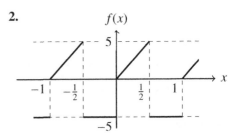

$f(x)$

The graph shows that the period of $f(x)$ is 1. Therefore

$$f(x + 1) = f(x)$$

A period is made up of 1 horizontal line and 1 slanted line.

Line 1 Line 1 is simply an horizontal line:

$$f(x) = -5 \quad (-\frac{1}{2} < x < 0)$$

Line 2 Line 2 passes through two points:

$$(0, 0) \quad \text{and} \quad (\frac{1}{2}, 5)$$

Thus

$$y = \left(\frac{y_1 - y_2}{x_1 - x_2}\right)(x - x_1) + y_1$$

$$= \left(\frac{0 - 5}{0 - \frac{1}{2}}\right)(x - 0) + 0$$

$$= \left(\frac{5}{\frac{1}{2}}\right)x$$

$$= 10x$$

Thus

$$f(x) = 10x \quad (0 < x < \frac{1}{2})$$

Therefore

$$f(x) = \begin{cases} -5 & -\frac{1}{2} < x < 0 \\ 10x & 0 < x < \frac{1}{2} \end{cases}$$

$$f(x + 1) = f(x)$$

3.

$$f(x) = \begin{cases} -1 & -\pi < x < -\frac{\pi}{2} \\ \frac{2x}{\pi} & -\frac{\pi}{2} < x < \frac{\pi}{2} \quad (1) \\ 1 & \frac{\pi}{2} < x < \pi \end{cases}$$

$$f(x + 2\pi) = f(x) \quad (2)$$

First, we shall draw the piecewise function only using (1)

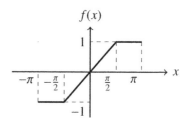

Next, we shall apply (2) by repeating the graph left and right.

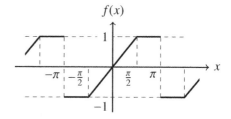

Chapter 3

Exercise 1

1.

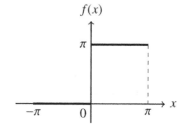

Solution for analytical definition
The analytical definition of $f(x)$ is

$$f(x) = \begin{cases} 0 & -\pi < x < 0 \\ \pi & 0 < x < \pi \end{cases}$$

$$f(x + 2\pi) = f(x)$$

Solution for a_0 The area under a single period of the graph of $f(x)$ is the shaded area shown below.

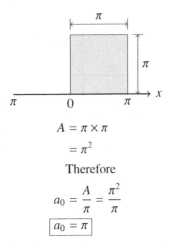

$$A = \pi \times \pi$$
$$= \pi^2$$

Therefore

$$a_0 = \frac{A}{\pi} = \frac{\pi^2}{\pi}$$

$$\boxed{a_0 = \pi}$$

Solution for a_n

$a_n\pi$ = sum of cases

$$= 0(-\pi, 0) + \pi(0, \pi)$$
$$= \pi(0, \pi) \qquad\qquad \text{(RP)}$$

apply Rule 3a

$$= \frac{\pi}{n} \sin n\pi$$
$$a_n\pi = 0$$

$$\boxed{a_n = 0}$$

Solution for b_n Continuing from the reuse point,

$$b_n\pi = \pi(0, \pi)$$

apply Rule 3b

$$b_n\pi = \frac{\pi}{n}(1 - \cos n\pi)$$
$$b_n = \frac{1}{n}(1 - \cos n\pi)$$

Therefore

$$b_n = \begin{cases} \dfrac{2}{n} & n \text{ is odd} \\[2mm] 0 & n \text{ is even} \end{cases}$$

Solution for Fourier series

$$f(x) = \frac{a_0}{2} + \sum_{n=1}^{\infty} a_n \cos nx + \sum_{n=1}^{\infty} b_n \sin nx$$

$$= \frac{\pi}{2} + 0(\cos x + \cos 2x + \cdots)$$
$$\qquad + b_n(\sin x + \sin 2x + \cdots)$$

$$= \frac{\pi}{2} + \frac{2}{n}(\sin x + \sin 3x + \cdots)$$
$$\qquad + 0(\sin 2x + \sin 4x + \cdots)$$

$$= \frac{\pi}{2} + \frac{2}{n}(\sin x + \sin 3x + \cdots)$$

$$\boxed{f(x) = \frac{\pi}{2} + 2(\sin x + \frac{1}{3}\sin 3x + \cdots)}$$

2.

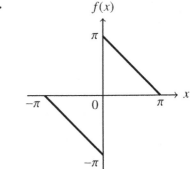

Solution for analytical definition

The analytical definition of $f(x)$ is

$$f(x) = \begin{cases} -x - \pi & -\pi < x < 0 \\ -x + \pi & 0 < x < \pi \end{cases}$$

$$f(x + 2\pi) = f(x)$$

Solution for a_0 The area under one period of the graph of $f(x)$ is the shaded area shown below.

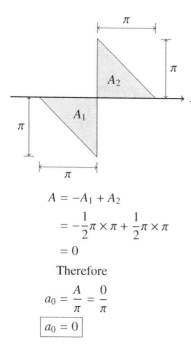

$$A = -A_1 + A_2$$

$$= -\frac{1}{2}\pi \times \pi + \frac{1}{2}\pi \times \pi$$

$$= 0$$

Therefore

$$a_0 = \frac{A}{\pi} = \frac{0}{\pi}$$

$$\boxed{a_0 = 0}$$

Solution for a_n

$a_n\pi$ = sum of cases

$$= (-x - \pi)(-\pi, 0) + (-x + \pi)(0, \pi)$$

$$= -x(-\pi, 0) - \pi(-\pi, 0) - x(0, \pi)$$

$$+ \pi(0, \pi)$$

$$= x(0, -\pi) + \pi(0, -\pi) - x(0, \pi)$$

$$+ \pi(0, \pi)$$

$$= x(0, -\pi) - x(0, \pi) + \pi(0, -\pi)$$

$$+ \pi(0, \pi) \qquad \text{(RP)}$$

apply Rules 3a and 4a

$$= -\frac{\pi}{n}\sin n(-\pi) + \frac{1}{n^2}[\cos n(-\pi) - 1]$$

$$- [\frac{\pi}{n}\sin n\pi + \frac{1}{n^2}(\cos n\pi - 1)]$$

$$+ \frac{\pi}{n}\sin n(-\pi) + \frac{\pi}{n}\sin n\pi$$

$$= 0 + \frac{1}{n^2}(\cos n\pi - 1)$$

$$- [0 + \frac{1}{n^2}(\cos n\pi - 1)] + 0 + 0$$

$$= \frac{1}{n^2}(\cos n\pi - 1) - \frac{1}{n^2}(\cos n\pi - 1)$$

$$a_n\pi = 0$$

$$\boxed{a_n = 0}$$

Solution for b_n Continuing from (RP),

$$b_n\pi = x(0, -\pi) - x(0, \pi)$$

$$+ \pi(0, -\pi) + \pi(0, \pi)$$

apply Rule 3b and 4b

$$= -\frac{(-\pi)}{n}\cos n(-\pi) + \frac{1}{n^2}\sin n(-\pi)$$

$$- [\frac{(-\pi)}{n}\cos n\pi + \frac{1}{n^2}\sin n\pi]$$

$$+ \frac{\pi}{n}[1 - \cos n(-\pi)] + \frac{\pi}{n}(1 - \cos n\pi)$$

$$= \frac{\pi}{n}\cos n\pi + 0 + \frac{\pi}{n}\cos n\pi + 0$$

$$+ \frac{\pi}{n}(1 - \cos n\pi) + \frac{\pi}{n}(1 - \cos n\pi)$$

$$= \frac{2\pi}{n}\cos n\pi + \frac{2\pi}{n}(1 - \cos n\pi)$$

$$b_n\pi = \frac{2\pi}{n}(\cos n\pi + 1 - \cos n\pi)$$

$$\boxed{b_n = \frac{2}{n}}$$

Solution for Fourier series

$$f(x) = \frac{a_0}{2} + \sum_{n=1}^{\infty}a_n\cos nx + \sum_{n=1}^{\infty}b_n\sin nx$$

$$= 0 + 0(\cos x + \cos 2x + \cos 3x + \cdots)$$

$$+ b_n(\sin x + \sin 2x + \sin 3x + \cdots)$$

$$= \frac{2}{n}(\sin x + \sin 2x + \sin 3x + \cdots)$$

$$\boxed{f(x) = 2(\sin x + \frac{1}{2}\sin 2x + \frac{1}{3}\sin 3x + \cdots)}$$

3.

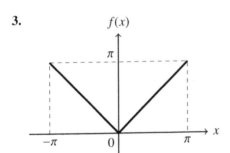

$f(x)$

Solution for analytical definition

The analytical definition of $f(x)$ is

$$f(x) = \begin{cases} -x & -\pi < x < 0 \\ x & 0 < x < \pi \end{cases}$$

$$f(x + 2\pi) = f(x)$$

Solution for a_0 The area under one period of the graph of $f(x)$ is the shaded area shown below.

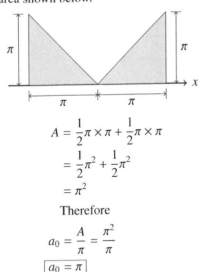

$$A = \frac{1}{2}\pi \times \pi + \frac{1}{2}\pi \times \pi$$

$$= \frac{1}{2}\pi^2 + \frac{1}{2}\pi^2$$

$$= \pi^2$$

Therefore

$$a_0 = \frac{A}{\pi} = \frac{\pi^2}{\pi}$$

$$\boxed{a_0 = \pi}$$

Solution for a_n

$a_n\pi$ = sum of cases

$$= -x(-\pi, 0) + x(0, \pi)$$

$$= x(0, -\pi) + x(0, \pi) \qquad \text{(RP)}$$

apply Rule 4a

$$= -\frac{\pi}{n}\sin n(-\pi) + \frac{1}{n^2}[\cos n(-\pi) - 1]$$

$$+ \frac{\pi}{n}\sin n\pi + \frac{1}{n^2}(\cos n\pi - 1)$$

$$= 0 + \frac{1}{n^2}(\cos n\pi - 1) + 0$$

$$+ \frac{1}{n^2}(\cos n\pi - 1)$$

$$= \frac{1}{n^2}(\cos n\pi - 1) + \frac{1}{n^2}(\cos n\pi - 1)$$

$$a_n\pi = \frac{2}{n^2}(\cos n\pi - 1)$$

$$a_n = \frac{2}{\pi n^2}(\cos n\pi - 1)$$

Therefore

$$a_n = \begin{cases} -\dfrac{4}{\pi n^2} & n \text{ is odd} \\[2mm] 0 & n \text{ is even} \end{cases}$$

Solution for b_n Continuing from reuse point,

$$b_n\pi = x(0, -\pi) + x(0, \pi)$$

apply Rule 4b

$$= -\frac{(-\pi)}{n}\cos n(-\pi) + \frac{1}{n^2}\sin n(-\pi)$$

$$- \frac{\pi}{n}\cos n\pi + \frac{1}{n^2}\sin n\pi$$

$$= \frac{\pi}{n}\cos n\pi + 0 - \frac{\pi}{n}\cos n\pi + 0$$

$$= \frac{\pi}{n}\cos n\pi - \frac{\pi}{n}\cos n\pi$$

$$b_n\pi = 0$$

$$\boxed{b_n = 0}$$

Solution for Fourier series

$$f(x) = \frac{a_0}{2} + \sum_{n=1}^{\infty} a_n \cos nx + \sum_{n=1}^{\infty} b_n \sin nx$$

$$= \frac{\pi}{2} + a_n\left(\cos x + \cos 3x + \cdots\right)$$

$$+ a_n\left(\cos 2x + \cos 4x + \cdots\right) + 0$$

$$= \frac{\pi}{2} - \frac{4}{\pi n^2}\left(\cos x + \cos 3x + \cdots\right)$$

$$+ 0\left(\cos 2x + \cos 4x + \cdots\right)$$

$$\boxed{f(x) = \frac{\pi}{2} - \frac{4}{\pi}\left(\cos x + \frac{1}{3^2}\cos 3x + \cdots\right)}$$

4. $f(x)$

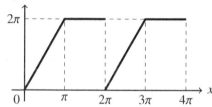

Solution for analytical definition

The analytical definition of $f(x)$ is

$$f(x) = \begin{cases} 2x & 0 < x < \pi \\ 2\pi & \pi < x < 2\pi \end{cases}$$

$$f(x + 2\pi) = f(x)$$

Solution for a_0 The area under one period of the graph of $f(x)$ is the shaded area shown below.

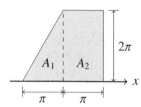

$$A = A_1 + A_2$$

$$A = \frac{1}{2}\pi \times 2\pi + \pi \times 2\pi$$

$$= \frac{2\pi^2}{2} + 2\pi^2$$

$$= \pi^2 + 2\pi^2$$

$$A = 3\pi^2$$

Therefore

$$a_0 = \frac{A}{\pi} = \frac{3\pi^2}{\pi}$$

$$\boxed{a_0 = 3\pi}$$

Solution for a_n

$$a_n\pi = \text{sum of cases}$$

$$= 2x(0, \pi) + 2\pi(\pi, 2\pi)$$

$$a_n\frac{\pi}{2} = x(0, \pi) + \pi(\pi, 2\pi)$$

$$= x(0, \pi) + \pi(\pi, 0) + \pi(0, 2\pi)$$

$$= x(0, \pi) - \pi(0, \pi)$$

$$+ \pi(0, 2\pi) \qquad \text{(RP)}$$

apply Rules 3a and 4a

$$= \frac{\pi}{n}\sin n\pi + \frac{1}{n^2}(\cos n\pi - 1)$$

$$- \frac{\pi}{n}\sin n\pi + \frac{\pi}{n}\sin n(2\pi)$$

$$a_n\frac{\pi}{2} = 0 + \frac{1}{n^2}(\cos n\pi - 1) - 0 + 0$$

$$a_n\pi = \frac{2}{n^2}(\cos n\pi - 1)$$

$$a_n = \frac{2}{\pi n^2}(\cos n\pi - 1)$$

Therefore

$$\boxed{a_n = \begin{cases} -\dfrac{4}{\pi n^2} & n \text{ is odd} \\ 0 & n \text{ is even} \end{cases}}$$

Solution for b_n Continuing from (RP),

$$b_n \frac{\pi}{2} = x(0, \pi) - \pi(0, \pi) + \pi(0, 2\pi)$$

apply Rules 3b and 4b

$$= -\frac{\pi}{n} \cos n\pi + \frac{1}{n^2} \sin n\pi$$

$$- \frac{\pi}{n}(1 - \cos n\pi) + \frac{\pi}{n}(1 - \cos 2n\pi)$$

$$= -\frac{\pi}{n} \cos n\pi + 0$$

$$- \frac{\pi}{n}(1 - \cos n\pi) + \frac{\pi}{n}(1 - 1)$$

$$= -\frac{\pi}{n} \cos n\pi - \frac{\pi}{n}(1 - \cos n\pi) + 0$$

$$= \frac{\pi}{n}(-\cos n\pi - 1 + \cos n\pi)$$

$$b_n \frac{\pi}{2} = \frac{\pi}{n}(-1)$$

$$\boxed{b_n = -\frac{2}{n}}$$

Solution for Fourier series

$$f(x) = \frac{a_0}{2} + \sum_{n=1}^{\infty} a_n \cos nx + \sum_{n=1}^{\infty} b_n \sin nx$$

$$= \frac{3\pi}{2} + a_n\left(\cos x + \cos 3x + \cdots\right)$$

$$+ a_n\left(\cos 2x + \cos 4x + \cdots\right)$$

$$+ b_n\left(\sin x + \sin 2x + \cdots\right)$$

$$= \frac{3\pi}{2} - \frac{4}{\pi n^2}\left(\cos x + \cos 3x + \cdots\right) + 0$$

$$- \frac{2}{n}\left(\sin x + \sin 2x + \cdots\right)$$

$$\boxed{\begin{aligned} f(x) = \frac{3\pi}{2} - \frac{4}{\pi}\Big(\cos x + \frac{1}{3^2}\cos 3x + \cdots\Big) \\ - 2\Big(\sin x + \frac{1}{2}\sin 2x + \cdots\Big) \end{aligned}}$$

Exercise 2

1.

$$f(x) = x + \pi \quad -\pi < x < \pi$$

$$f(x + 2\pi) = f(x)$$

Solution for graph of $f(x)$ The graph of $f(x)$ can be drawn as

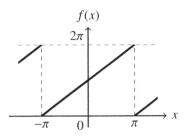

Solution for a_0 The area under one period of the graph of $f(x)$ is the shaded area shown below.

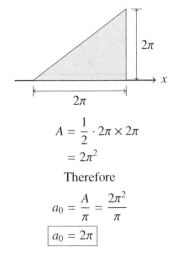

$$A = \frac{1}{2} \cdot 2\pi \times 2\pi$$

$$= 2\pi^2$$

Therefore

$$a_0 = \frac{A}{\pi} = \frac{2\pi^2}{\pi}$$

$$\boxed{a_0 = 2\pi}$$

Solution for a_n

$$\begin{aligned} a_n \pi &= \text{sum of cases} \\ &= (x + \pi)(-\pi, \pi) \\ &= x(-\pi, \pi) + \pi(-\pi, \pi) \\ &= x(-\pi, 0) + x(0, \pi) \\ &\quad + \pi(-\pi, 0) + \pi(0, \pi) \\ &= -x(0, -\pi) + x(0, \pi) \\ &\quad - \pi(0, -\pi) + \pi(0, \pi) \end{aligned}$$

(RP)

apply Rules 3a and 4a

$$= -\left[\frac{(-\pi)}{n}\sin n(-\pi)\right.$$

$$\left. + \frac{1}{n^2}\big(\cos n(-\pi) - 1\big)\right]$$

$$+ \left[\frac{\pi}{n}\sin n\pi + \frac{1}{n^2}(\cos n\pi - 1)\right]$$

$$- \frac{\pi}{n}\sin n(-\pi) + \frac{\pi}{n}\sin n(\pi)$$

$$= -\left[0 + \frac{1}{n^2}(\cos n\pi - 1)\right]$$

$$+ \left[0 + \frac{1}{n^2}(\cos n\pi - 1)\right] - 0 + 0$$

$$a_n\pi = -\frac{1}{n^2}(\cos n\pi - 1) + \frac{1}{n^2}(\cos n\pi - 1)$$

$$\boxed{a_n = 0}$$

Solution for b_n   Continuing from (RP),

$$b_n\pi = -x(0, -\pi) + x(0, \pi)$$

$$- \pi(0, -\pi) + \pi(0, \pi)$$

apply Rules 3b and 4b

$$= -\left[-\frac{(-\pi)}{n}\cos n(-\pi) + \frac{1}{n^2}\sin n(-\pi)\right]$$

$$+ \left[-\frac{\pi}{n}\cos n\pi + \frac{1}{n^2}\sin n\pi\right]$$

$$- \frac{\pi}{n}\big[1 - \cos n(-\pi)\big] + \frac{\pi}{n}\big[1 - \cos n\pi\big]$$

$$= -\left[\frac{\pi}{n}\cos n\pi + 0\right]$$

$$+ \left[-\frac{\pi}{n}\cos n\pi + 0\right]$$

$$- \frac{\pi}{n}\big[1 - \cos n\pi\big] + \frac{\pi}{n}\big[1 - \cos n\pi\big]$$

$$= -\frac{\pi}{n}\cos n\pi - \frac{\pi}{n}\cos n\pi + 0$$

$$b_n\pi = -\frac{2\pi}{n}\cos n\pi$$

$$b_n = -\frac{2}{n}\cos n\pi$$

Therefore

$$\boxed{b_n = \begin{cases} \dfrac{2}{n} & n \text{ is odd} \\[2mm] -\dfrac{2}{n} & n \text{ is even} \end{cases}}$$

Solution for Fourier series

$$f(x) = \frac{a_0}{2} + \sum_{n=1}^{\infty} a_n \cos nx + \sum_{n=1}^{\infty} b_n \sin nx$$

$$= \frac{2\pi}{2} + 0 + b_n\big(\sin x + \sin 2x + \cdots\big)$$

$$= \pi + b_n\big(\sin x + \sin 3x + \cdots\big)$$

$$+ b_n\big(\sin 2x + \sin 4x + \cdots\big)$$

$$= \pi + \frac{2}{n}\big(\sin x + \sin 3x + \cdots\big)$$

$$- \frac{2}{n}\big(\sin 2x + \sin 4x + \cdots\big)$$

$$= \pi + 2\big(\sin x + \frac{1}{3}\sin 3x + \cdots\big)$$

$$- 2\big(\frac{1}{2}\sin 2x + \frac{1}{4}\sin 4x + \cdots\big)$$

$$\boxed{f(x) = \pi + 2\big(\sin x - \frac{1}{2}\sin 2x + \cdots\big)}$$

2.

$$f(x) = \begin{cases} 0 & -\pi < x < 0 \\ 1 & 0 < x < \pi \end{cases}$$

Solution for graph of $f(x)$ The graph of $f(x)$ can be drawn as

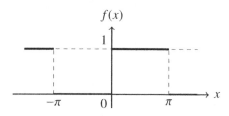

Solution for a_0   The area under one period of the graph of $f(x)$ is the shaded area shown below.

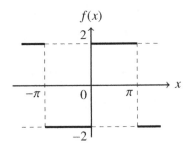

$$A = \pi \times 1 = \pi$$

Therefore

$$a_0 = \frac{A}{\pi} = \frac{\pi}{\pi}$$

$$\boxed{a_0 = 1}$$

Solution for a_n

$a_n\pi$ = sum of cases

$$= 0(-\pi, 0) + 1(0, \pi)$$

$$= (0, \pi) \hspace{2cm} \text{(RP)}$$

apply Rule 3a

$$= \frac{1}{n}\sin n\pi$$

$a_n\pi = 0$

$$\boxed{a_n = 0}$$

Solution for b_n   Continuing from (RP),

$b_n\pi = (0, \pi)$

apply Rule 3b

$$= \frac{1}{n}(1 - \cos n\pi)$$

$$b_n = \frac{1}{n\pi}(1 - \cos n\pi)$$

Therefore

$$\boxed{b_n = \begin{cases} \dfrac{2}{n\pi} & n \text{ is odd} \\ 0 & n \text{ is even} \end{cases}}$$

Solution for Fourier series

$$f(x) = \frac{a_0}{2} + \sum_{n=1}^{\infty} a_n \cos nx + \sum_{n=1}^{\infty} b_n \sin nx$$

$$= \frac{1}{2} + 0 + b_n\Big(\sin x + \sin 2x + \cdots\Big)$$

$$= \frac{1}{2} + b_n\Big(\sin x + \sin 3x + \cdots\Big)$$

$$\hspace{1cm} + b_n\Big(\sin 2x + \sin 4x + \cdots\Big)$$

$$= \frac{1}{2} + \frac{2}{n\pi}\Big(\sin x + \sin 3x + \cdots\Big)$$

$$\hspace{1cm} + 0\Big(\sin 2x + \sin 4x + \cdots\Big)$$

$$\boxed{f(x) = \frac{1}{2} + \frac{2}{\pi}\Big(\sin x + \frac{1}{3}\sin 3x + \cdots\Big)}$$

3.

$$f(x) = \begin{cases} -2 & -\pi < x < 0 \\ 2 & 0 < x < \pi \end{cases}$$

Solution for graph of $f(x)$ The graph of $f(x)$ can be drawn as

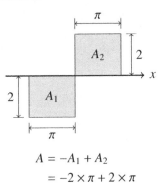

Solution for a_0   The area under one period of the graph of $f(x)$ is the shaded area shown below.

$$A = -A_1 + A_2$$

$$= -2 \times \pi + 2 \times \pi$$

$$= 0$$

Therefore

$$a_0 = \frac{A}{\pi} = \frac{0}{\pi}$$

$$\boxed{a_0 = 0}$$

Solution for a_n

$a_n \pi$ = sum of cases

$$= -2(-\pi, 0) + 2(0, \pi)$$

$$= 2(0, -\pi) + 2(0, \pi) \qquad \text{(RP)}$$

apply Rule 3a

$$= \frac{2}{n} \sin n(-\pi) + \frac{2}{n} \sin n\pi$$

$$= 0 + 0$$

$a_n \pi = 0$

$$\boxed{a_n = 0}$$

Solution for b_n   Continuing from (RP),

$$b_n \pi = 2(0, -\pi) + 2(0, \pi)$$

apply Rule 3b

$$= \frac{2}{n}[1 - \cos n(-\pi)] + \frac{2}{n}(1 - \cos n\pi)$$

$$= \frac{2}{n}(1 - \cos n\pi) + \frac{2}{n}(1 - \cos n\pi)$$

$$b_n \pi = \frac{4}{n}(1 - \cos n\pi)$$

$$b_n = \frac{4}{n\pi}(1 - \cos n\pi)$$

Therefore

$$\boxed{b_n = \begin{cases} \dfrac{8}{n\pi} & n \text{ is odd} \\ 0 & n \text{ is even} \end{cases}}$$

Solution for Fourier series

$$f(x) = \frac{a_0}{2} + \sum_{n=1}^{\infty} a_n \cos nx + \sum_{n=1}^{\infty} b_n \sin nx$$

$$= \frac{0}{2} + 0 + b_n\left(\sin x + \sin 2x + \cdots\right)$$

$$= b_n\left(\sin x + \sin 3x + \cdots\right)$$

$$+ b_n\left(\sin 2x + \sin 4x + \cdots\right)$$

$$= \frac{8}{n\pi}\left(\sin x + \sin 3x + \cdots\right)$$

$$+ 0\left(\sin 2x + \sin 4x + \cdots\right)$$

$$= \frac{8}{\pi}\left(\sin x + \frac{1}{3}\sin 3x + \cdots\right) + 0$$

$$\boxed{f(x) = \frac{8}{\pi}\left(\sin x + \frac{1}{3}\sin 3x + \cdots\right)}$$

4.

$$f(x) = \begin{cases} 1 + \dfrac{2x}{\pi} & -\pi < x < 0 \\ 1 - \dfrac{2x}{\pi} & 0 < x < \pi \end{cases}$$

Solution for graph of $f(x)$ The graph of $f(x)$ can be drawn as

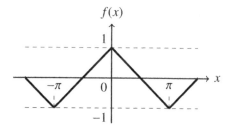

Solution for a_0   The area under one period of the graph of $f(x)$ is the shaded area shown below.

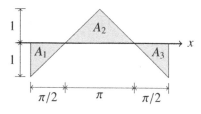

$$A = -A_1 + A_2 - A_3$$

$$= -\frac{1}{2}\left(\frac{\pi}{2} \times 1\right) + \frac{1}{2}\left(\pi \times 1\right) - \frac{1}{2}\left(\frac{\pi}{2} \times 1\right)$$

$$= -\frac{\pi}{4} + \frac{\pi}{2} - \frac{\pi}{4}$$

$$= 0$$

Therefore

$$a_0 = \frac{A}{\pi} = \frac{0}{\pi}$$

$$\boxed{a_0 = 0}$$

Solution for a_n

$a_n\pi$

= sum of cases

$$= \left(1 + \frac{2x}{\pi}\right)(-\pi, 0) + \left(1 - \frac{2x}{\pi}\right)(0, \pi)$$

$$= (-\pi, 0) + \frac{2x}{\pi}(-\pi, 0)$$

$$+ (0, \pi) - \frac{2x}{\pi}(0, \pi)$$

$$= -(0, -\pi) - \frac{2x}{\pi}(0, -\pi)$$

$$+ (0, \pi) - \frac{2x}{\pi}(0, \pi) \qquad \text{(RP)}$$

apply Rules 3a and 4a

$$= -\frac{1}{n}\sin n(-\pi)$$

$$- \frac{2}{\pi}\left[\frac{-\pi}{n}\sin n(-\pi) + \frac{1}{n^2}\Big(\cos n(-\pi) - 1\Big)\right]$$

$$+ \frac{1}{n}\sin n(\pi)$$

$$- \frac{2}{\pi}\left[\frac{\pi}{n}\sin n\pi + \frac{1}{n^2}\Big(\cos n\pi - 1\Big)\right]$$

$$= -0 - \frac{2}{\pi}\left[0 + \frac{1}{n^2}\Big(\cos n\pi - 1\Big)\right]$$

$$+ 0 - \frac{2}{\pi}\left[0 + \frac{1}{n^2}\Big(\cos n\pi - 1\Big)\right]$$

$$= -\frac{2}{\pi n^2}\Big(\cos n\pi - 1\Big) - \frac{2}{\pi n^2}\Big(\cos n\pi - 1\Big)$$

$$= -\frac{4}{\pi n^2}\Big(\cos n\pi - 1\Big)$$

Therefore

$$a_n = \begin{cases} \dfrac{8}{\pi^2 n^2} & n \text{ is odd} \\ 0 & n \text{ is even} \end{cases}$$

Solution for b_n Continuing from (RP),

$$b_n\pi = -(0, -\pi) - \frac{2x}{\pi}(0, -\pi)$$

$$+ (0, \pi) - \frac{2x}{\pi}(0, \pi)$$

apply Rules 3b and 4b

$$= -\frac{1}{n}[1 - \cos n(-\pi)]$$

$$- \frac{2}{\pi}\left[-\frac{-\pi}{n}\cos n(-\pi) + \frac{1}{n^2}\sin n(-\pi)\right]$$

$$+ \frac{1}{n}(1 - \cos n\pi)$$

$$- \frac{2}{\pi}\left[-\frac{\pi}{n}\cos n\pi + \frac{1}{n^2}\sin n\pi\right]$$

$$= -\frac{1}{n}(1 - \cos n\pi) - \frac{2}{\pi}\left[\frac{\pi}{n}\cos n\pi + 0\right]$$

$$+ \frac{1}{n}(1 - \cos n\pi) - \frac{2}{\pi}\left[-\frac{\pi}{n}\cos n\pi + 0\right]$$

$$= -\frac{1}{n}(1 - \cos n\pi) - \frac{2}{n}\cos n\pi$$

$$+ \frac{1}{n}(1 - \cos n\pi) + \frac{2}{n}\cos n\pi$$

$b_n\pi = 0$

$$\boxed{b_n = 0}$$

Solution for Fourier series

$$f(x) = \frac{a_0}{2} + \sum_{n=1}^{\infty} a_n\cos nx + \sum_{n=1}^{\infty} b_n\sin nx$$

$$= \frac{0}{2} + a_n\Big(\cos x + \cos 2x + \cdots\Big) + 0$$

$$= 0 + a_n\Big(\cos x + \cos 3x + \cdots\Big)$$

$$+ a_n\Big(\cos 2x + \cos 4x + \cdots\Big)$$

$$= \frac{8}{\pi^2 n^2}\Big(\cos x + \cos 3x + \cdots\Big)$$

$$+ 0\Big(\cos 2x + \cos 4x + \cdots\Big)$$

$$\boxed{f(x) = \frac{8}{\pi^2}\left(\cos x + \frac{1}{3^2}\cos 3x + \cdots\right)}$$

5.

$$f(x) = \begin{cases} 0 & -\pi < x < 0 \\ 1 & 0 < x < \dfrac{\pi}{2} \\ -1 & \dfrac{\pi}{2} < x < \pi \end{cases}$$

Solution for graph of $f(x)$ The graph of $f(x)$ can be drawn as

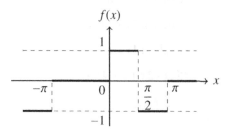

Solution for a_0 The area under one period of the graph of $f(x)$ is the shaded area shown below.

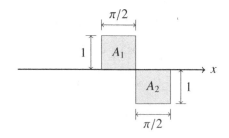

$$A = A_1 - A_2$$

$$= \frac{\pi}{2} \times 1 - \frac{\pi}{2} \times 1$$

$$= 0$$

Therefore

$$a_0 = \frac{A}{\pi} = \frac{0}{\pi}$$

$$\boxed{a_0 = 0}$$

Solution for a_n

$a_n\pi$ = sum of cases

$$= 0(-\pi, 0) + 1(0, \frac{\pi}{2}) - 1(\frac{\pi}{2}, \pi)$$

$$= (0, \frac{\pi}{2}) - (\frac{\pi}{2}, \pi)$$

$$= (0, \frac{\pi}{2}) - (\frac{\pi}{2}, 0) - (0, \pi)$$

$$= (0, \frac{\pi}{2}) + (0, \frac{\pi}{2}) - (0, \pi)$$

$$= 2(0, \frac{\pi}{2}) - (0, \pi) \qquad \text{(RP)}$$

apply Rule 3a

$$= \frac{2}{n} \sin n\left(\frac{\pi}{2}\right) - \frac{1}{n} \sin n\pi$$

$$a_n\pi = \frac{2}{n} \sin \frac{n\pi}{2} - 0$$

$$a_n = \frac{2}{n\pi} \sin \frac{n\pi}{2}$$

Recall that

$$\sin \frac{n\pi}{2} = \begin{cases} 0 & n \text{ is even} \\ 1 & n = 1, 5, 9, \ldots \\ -1 & n = 3, 7, 11, \ldots \end{cases}$$

Therefore

$$\boxed{a_n = \begin{cases} 0 & n \text{ is even} \\ \dfrac{2}{n\pi} & n = 1, 5, 9, \ldots \\ -\dfrac{2}{n\pi} & n = 3, 7, 11, \ldots \end{cases}}$$

Solution for b_n Continuing from (RP),

$$b_n\pi = 2(0, \frac{\pi}{2}) - (0, \pi)$$

apply Rule 3b

$$= \frac{2}{n}\left[1 - \cos n\left(\frac{\pi}{2}\right)\right] - \frac{1}{n}\left[1 - \cos n(\pi)\right]$$

$$= \frac{2}{n}\left(1 - \cos \frac{n\pi}{2}\right) - \frac{1}{n}\left(1 - \cos n\pi\right)$$

$$= \frac{1}{n}\left(2 - 2\cos \frac{n\pi}{2} - 1 + \cos n\pi\right)$$

$$b_n\pi = \frac{1}{n}\left(1 - 2\cos \frac{n\pi}{2} + \cos n\pi\right)$$

$$b_n = \frac{1}{n\pi}\left(\cos n\pi - 2\cos \frac{n\pi}{2} + 1\right)$$

Recall that

$$\cos n\pi = \begin{cases} -1 & n \text{ is odd} \\ 1 & n \text{ is even} \end{cases}$$

and

$$\cos\frac{n\pi}{2} = \begin{cases} 0 & n \text{ is odd} \\ -1 & n = 2, 6, 10, \ldots \\ 1 & n = 4, 8, 12, \ldots \end{cases}$$

When n is odd

$$b_n = \frac{1}{n\pi}(-1 - 2(0) + 1)$$

$$= \frac{1}{n\pi}(-1 + 1)$$

$$= 0$$

When $n = 2, 6, 10, \ldots$

$$b_n = \frac{1}{n\pi}(1 - 2(-1) + 1)$$

$$= \frac{1}{n\pi}(1 + 2 + 1)$$

$$= \frac{4}{n\pi}$$

When $n = 4, 8, 12, \ldots$

$$b_n = \frac{1}{n\pi}(1 - 2(1) + 1)$$

$$= \frac{1}{n\pi}(-1 + 1)$$

$$= 0$$

Therefore

$$b_n = \begin{cases} 0 & n \text{ is odd} \\ \dfrac{4}{n\pi} & n = 2, 6, 10, \ldots \\ 0 & n = 4, 8, 12, \ldots \end{cases}$$

Solution for Fourier series

$$f(x) = \frac{a_0}{2} + \sum_{n=1}^{\infty} a_n \cos nx + \sum_{n=1}^{\infty} b_n \sin nx$$

$$= 0 + a_n\left(\cos x + \cos 2x + \cdots\right)$$

$$+ b_n\left(\sin x + \sin 2x + \cdots\right)$$

$$= a_n\left(\cos 2x + \cos 4x + \cdots\right)$$

$$+ a_n\left(\cos x + \cos 5x + \cdots\right)$$

$$+ a_n\left(\cos 3x + \cos 7x + \cdots\right)$$

$$+ b_n\left(\sin x + \sin 3x + \cdots\right)$$

$$+ b_n\left(\sin 2x + \sin 6x + \cdots\right)$$

$$+ b_n\left(\sin 4x + \sin 8x + \cdots\right)$$

$$= 0\left(\cos 2x + \cos 4x + \cdots\right)$$

$$+ \frac{2}{n\pi}\left(\cos x + \cos 5x + \cdots\right)$$

$$- \frac{2}{n\pi}\left(\cos 3x + \cos 7x + \cdots\right)$$

$$+ 0\left(\sin x + \sin 3x + \cdots\right)$$

$$+ \frac{4}{n\pi}\left(\sin 2x + \sin 6x + \cdots\right)$$

$$+ 0\left(\sin 4x + \sin 8x + \cdots\right)$$

$$= 0 + \frac{2}{n\pi}\left(\cos x + \cos 5x + \cdots\right)$$

$$- \frac{2}{n\pi}\left(\cos 3x + \cos 7x + \cdots\right) + 0$$

$$+ \frac{4}{n\pi}\left(\sin 2x + \sin 6x + \cdots\right) + 0$$

$$= \frac{2}{\pi}\left(\cos x + \frac{1}{5}\cos 5x + \cdots\right)$$

$$- \frac{2}{\pi}\left(\frac{1}{3}\cos 3x + \frac{1}{7}\cos 7x + \cdots\right)$$

$$+ \frac{4}{\pi}\left(\frac{1}{2}\sin 2x + \frac{1}{6}\sin 6x + \cdots\right)$$

$$\boxed{f(x) = \frac{2}{\pi}\left(\cos x - \frac{1}{3}\cos 3x + \cdots\right) \\ + \frac{4}{\pi}\left(\frac{1}{2}\sin 2x + \frac{1}{6}\sin 6x + \cdots\right)}$$

6.

$$f(x) = \begin{cases} 0 & -\pi < x < 0 \\ x & 0 < x < \dfrac{\pi}{2} \\ \pi - x & \dfrac{\pi}{2} < x < \pi \end{cases}$$

Solution for graph of $f(x)$ The graph of $f(x)$ can be drawn as

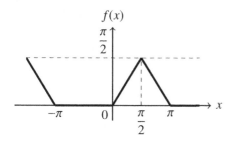

$f(x)$

Solution for a_0 The area under one period of the graph of $f(x)$ is the shaded area shown below.

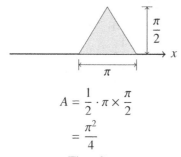

$$A = \frac{1}{2} \cdot \pi \times \frac{\pi}{2}$$

$$= \frac{\pi^2}{4}$$

Therefore

$$a_0 = \frac{A}{\pi} = \frac{\pi^2/4}{\pi}$$

$$\boxed{a_0 = \frac{\pi}{4}}$$

Solution for a_n

$a_n \pi$

$= $ sum of cases

$$= 0(-\pi, 0) + x(0, \frac{\pi}{2}) + (\pi - x)(\frac{\pi}{2}, \pi)$$

$$= 0 + x(0, \frac{\pi}{2}) + \pi(\frac{\pi}{2}, \pi) - x(\frac{\pi}{2}, \pi)$$

$$= x(0, \frac{\pi}{2}) + \pi(\frac{\pi}{2}, 0) + \pi(0, \pi)$$

$$- x(\frac{\pi}{2}, 0) - x(0, \pi)$$

$$= x(0, \frac{\pi}{2}) - \pi(0, \frac{\pi}{2}) + \pi(0, \pi)$$

$$+ x(0, \frac{\pi}{2}) - x(0, \pi)$$

$$= 2x(0, \frac{\pi}{2}) - x(0, \pi)$$

$$- \pi(0, \frac{\pi}{2}) + \pi(0, \pi) \qquad \text{(RP)}$$

apply Rules 3a and 4a

$$= 2\left[\frac{\pi}{2n} \sin n\left(\frac{\pi}{2}\right) + \frac{1}{n^2}\left(\cos n\left(\frac{\pi}{2}\right) - 1\right)\right]$$

$$- \left[\frac{\pi}{n} \sin n\pi + \frac{1}{n^2}\left(\cos n\pi - 1\right)\right]$$

$$- \frac{\pi}{n} \sin n\left(\frac{\pi}{2}\right) + \frac{\pi}{n} \sin n\pi$$

$$= 2\left[\frac{\pi}{2n} \sin \frac{n\pi}{2} + \frac{1}{n^2}\left(\cos \frac{n\pi}{2} - 1\right)\right]$$

$$- \left[0 + \frac{1}{n^2}\left(\cos n\pi - 1\right)\right]$$

$$- \frac{\pi}{n} \sin \frac{n\pi}{2} + 0$$

$$= \frac{\pi}{n} \sin \frac{n\pi}{2} + \frac{2}{n^2}\left(\cos \frac{n\pi}{2} - 1\right)$$

$$- \frac{1}{n^2}\left(\cos n\pi - 1\right) - \frac{\pi}{n} \sin \frac{n\pi}{2}$$

$$= \frac{2}{n^2}\left(\cos \frac{n\pi}{2} - 1\right) - \frac{1}{n^2}\left(\cos n\pi - 1\right)$$

$$= \frac{1}{n^2}\left(2\cos \frac{n\pi}{2} - 2 - \cos n\pi + 1\right)$$

$$= \frac{1}{n^2}\left(2\cos \frac{n\pi}{2} - \cos n\pi - 1\right)$$

Therefore

$$a_n = \frac{1}{\pi n^2}\left(2\cos \frac{n\pi}{2} - \cos n\pi - 1\right)$$

Recall that

$$\cos n\pi = \begin{cases} -1 & n \text{ is odd} \\ 1 & n \text{ is even} \end{cases}$$

and

$$\cos \frac{n\pi}{2} = \begin{cases} 0 & n \text{ is odd} \\ -1 & n = 2, 6, 10, \ldots \\ 1 & n = 4, 8, 12, \ldots \end{cases}$$

When n is odd

$$a_n = \frac{1}{\pi n^2}\big(2(0) - (-1) - 1\big)$$

$$= \frac{1}{\pi n^2}(1 - 1)$$

$$= 0$$

When $n = 2, 6, 10, \ldots$

$$a_n = \frac{1}{\pi n^2}\big(2(-1) - (1) - 1\big)$$

$$= \frac{1}{\pi n^2}(-2 - 2)$$

$$= -\frac{4}{\pi n^2}$$

When $n = 4, 8, 12, \ldots$

$$a_n = \frac{1}{\pi n^2}\big(2(1) - (1) - 1\big)$$

$$= \frac{1}{\pi n^2}(2 - 2)$$

$$= 0$$

Therefore

$$a_n = \begin{cases} 0 & n \text{ is odd} \\ -\dfrac{4}{\pi n^2} & n = 2, 6, 10, \ldots \\ 0 & n = 4, 8, 12, \ldots \end{cases}$$

Solution for b_n   Continuing from (RP),

$$b_n\pi = 2x(0, \tfrac{\pi}{2}) - x(0, \pi)$$

$$- \pi(0, \tfrac{\pi}{2}) + \pi(0, \pi)$$

apply Rules 3b and 4b

$$= 2\Big[-\frac{\pi/2}{n}\cos\frac{n\pi}{2} + \frac{1}{n^2}\sin\frac{n\pi}{2} \Big]$$

$$- \Big[-\frac{\pi}{n}\cos n\pi + \frac{1}{n^2}\sin n\pi \Big]$$

$$- \frac{\pi}{n}\Big(1 - \cos\frac{n\pi}{2}\Big) + \frac{\pi}{n}\big(1 - \cos n\pi\big)$$

$$= -\frac{\pi}{n}\cos\frac{n\pi}{2} + \frac{2}{n^2}\sin\frac{n\pi}{2}$$

$$+ \frac{\pi}{n}\cos n\pi - 0$$

$$- \frac{\pi}{n}\Big(1 - \cos\frac{n\pi}{2}\Big) + \frac{\pi}{n}\big(1 - \cos n\pi\big)$$

$$= \frac{\pi}{n}\Big(-\cos\frac{n\pi}{2} + \cos n\pi - 1$$

$$+ \cos\frac{n\pi}{2} + 1 - \cos n\pi\Big)$$

$$+ \frac{2}{n^2}\sin\frac{n\pi}{2}$$

$$= \frac{2}{n^2}\sin\frac{n\pi}{2}$$

Thus

$$\boxed{\; b_n = \frac{2}{\pi n^2}\sin\frac{n\pi}{2} \;}$$

Recall that

$$\sin\frac{n\pi}{2} = \begin{cases} 0 & n \text{ is even} \\ 1 & n = 1, 5, 9, \ldots \\ -1 & n = 3, 7, 11, \ldots \end{cases}$$

Therefore

$$b_n = \begin{cases} 0 & n \text{ is even} \\ \dfrac{2}{\pi n^2} & n = 1, 5, 9, \ldots \\ -\dfrac{2}{\pi n^2} & n = 3, 7, 11, \ldots \end{cases}$$

Solution for Fourier series

$$f(x) = \frac{a_0}{2} + \sum_{n=1}^{\infty} a_n \cos nx + \sum_{n=1}^{\infty} b_n \sin nx$$

$$= \frac{\pi}{8} + a_n\big(\cos x + \cos 2x + \cdots\big)$$

$$+ b_n\big(\sin x + \sin 2x + \cdots\big)$$

$$= \frac{\pi}{8} + a_n\big(\cos x + \cos 3x + \cdots\big)$$

$$+ a_n\big(\cos 2x + \cos 6x + \cdots\big)$$

$$+ a_n\big(\cos 4x + \cos 8x + \cdots\big)$$

$$+ b_n\big(\sin 2x + \sin 4x + \cdots\big)$$

$$+ b_n\big(\sin x + \sin 5x + \cdots\big)$$

$$+ b_n\big(\sin 3x + \sin 7x + \cdots\big)$$

$$= \frac{\pi}{8} + 0\left(\cos x + \cos 3x + \cdots \right)$$

$$- \frac{4}{\pi n^2}\left(\cos 2x + \cos 6x + \cdots \right)$$

$$+ 0\left(\cos 4x + \cos 8x + \cdots \right)$$

$$+ 0\left(\sin 2x + \sin 4x + \cdots \right)$$

$$+ \frac{2}{\pi n^2}\left(\sin x + \sin 5x + \cdots \right)$$

$$- \frac{2}{\pi n^2}\left(\sin 3x + \sin 7x + \cdots \right)$$

$$= \frac{\pi}{8} - \frac{4}{\pi n^2}\left(\cos 2x + \cos 6x + \cdots \right)$$

$$+ \frac{2}{\pi n^2}\left(\sin x - \sin 3x + \cdots \right)$$

$$\boxed{\begin{aligned} f(x) &= \frac{\pi}{8} - \frac{4}{\pi}\left(\frac{1}{2^2} \cos 2x + \frac{1}{6^2} \cos 6x + \cdots \right) \\ &\quad + \frac{2}{\pi}\left(\sin x - \frac{1}{3^2} \sin 3x + \cdots \right) \end{aligned}}$$

Chapter 4

Exercise 1

1.

$$f(x) = 3 - 2x \qquad -\pi < x < \pi$$

Solution for a_0

$$a_0\pi = \text{sum of cases}$$
$$= (3 - 2x)(-\pi, \pi)$$
$$= 3(-\pi, \pi) - 2x(-\pi, \pi)$$
$$= 3(-\pi, 0) + 3(0, \pi)$$
$$\quad - 2x(-\pi, 0) - 2x(0, \pi)$$
$$= -3(0, -\pi) + 3(0, \pi)$$
$$\quad + 2x(0, -\pi) - 2x(0, \pi) \qquad \text{(RP)}$$

apply Rules 3c and 4c

$$= -3(-\pi) + 3(\pi) + 2 \cdot \frac{(-\pi)^2}{2} - 2 \cdot \frac{\pi^2}{2}$$
$$= 3\pi + 3\pi + \pi^2 - \pi^2$$

$$a_0\pi = 6\pi$$

$$\boxed{a_0 = 6}$$

Solution for a_n Continuing from the reuse point,

$$a_n\pi = -3(0, -\pi) + 3(0, \pi)$$
$$\quad + 2x(0, -\pi) - 2x(0, \pi)$$

apply Rules 3a and 4a

$$= -\frac{3}{n} \sin n(-\pi) + \frac{3}{n} \sin n\pi$$

$$+ 2\left[-\frac{\pi}{n} \sin n(-\pi) + \frac{1}{n^2}(\cos n(-\pi) - 1) \right]$$

$$- 2\left[\frac{\pi}{n} \sin n\pi + \frac{1}{n^2}(\cos n\pi - 1) \right]$$

$$= -0 + 0 + 2\left[-0 + \frac{1}{n^2}(\cos n\pi - 1) \right]$$

$$- 2\left[0 + \frac{1}{n^2}(\cos n\pi - 1) \right]$$

$$a_n\pi = \frac{2}{n^2}(\cos n\pi - 1) - \frac{2}{n^2}(\cos n\pi - 1)$$

$$\boxed{a_n = 0}$$

Solution for b_n Continuing from the reuse point,

$$b_n\pi = -3(0, -\pi) + 3(0, \pi)$$
$$\quad + 2x(0, -\pi) - 2x(0, \pi)$$

apply Rules 3b and 4b

$$= -\frac{3}{n}[1 - \cos n(-\pi)] + \frac{3}{n}(1 - \cos n\pi)$$

$$+ 2\left[-\frac{(-\pi)}{n} \cos n(-\pi) + \frac{1}{n^2} \sin n(-\pi) \right]$$

$$- 2\left[-\frac{\pi}{n} \cos n\pi + \frac{1}{n^2} \sin n\pi \right]$$

$$= -\frac{3}{n}(1 - \cos n\pi) + \frac{3}{n}(1 - \cos n\pi)$$

$$+ 2\left[\frac{\pi}{n} \cos n\pi + 0 \right]$$

$$- 2\left[-\frac{\pi}{n} \cos n\pi + 0 \right]$$

$$b_n\pi = 0 + \frac{2\pi}{n} \cos n\pi + \frac{2\pi}{n} \cos n\pi$$

$$b_n = \frac{4}{n} \cos n\pi$$

Therefore

$$b_n = \begin{cases} -\dfrac{4}{n} & n \text{ is odd} \\ \dfrac{4}{n} & n \text{ is even} \end{cases}$$

Solution for Fourier series

$$f(x) = \frac{a_0}{2} + \sum_{n=1}^{\infty} a_n \cos nx + \sum_{n=1}^{\infty} b_n \sin nx$$

$$= \frac{6}{2} + a_n \Big(\cos x + \cos 2x + \cdots \Big)$$

$$+ b_n \Big(\sin x + \sin 2x + \cdots \Big)$$

$$= 3 + 0 + b_n \Big(\sin x + \sin 3x + \cdots \Big)$$

$$+ b_n \Big(\sin 2x + \sin 4x + \cdots \Big)$$

$$= 3 - \frac{4}{n} \Big(\sin x + \sin 3x + \cdots \Big)$$

$$+ \frac{4}{n} \Big(\sin 2x + \sin 4x + \cdots \Big)$$

$$= 3 - 4 \Big(\sin x + \frac{1}{3} \sin 3x + \cdots \Big)$$

$$+ 4 \Big(\frac{1}{2} \sin 2x + \frac{1}{4} \sin 4x + \cdots \Big)$$

$$\boxed{f(x) = 3 - 4 \Big(\sin x - \frac{1}{2} \sin 2x + \cdots \Big)}$$

2.

$$f(x) = \begin{cases} -1 & -\pi < x < 0 \\ 2 & 0 < x < \pi \end{cases}$$

Solution for a_0

$$a_0 \pi = \text{sum of cases}$$

$$= -1(-\pi, 0) + 2(0, \pi)$$

$$= (0, -\pi) + 2(0, \pi) \qquad \text{(RP)}$$

apply Rule 3c

$$= (-\pi) + 2(\pi)$$

$$a_0 \pi = \pi$$

$$\boxed{a_0 = 1}$$

Solution for a_n   Continuing from the reuse point,

$$a_n \pi = (0, -\pi) + 2(0, \pi)$$

apply Rule 3a

$$= \frac{1}{n} \sin n(-\pi) + \frac{2}{n} \sin n\pi$$

$$a_n \pi = -0 + 0$$

$$\boxed{a_n = 0}$$

Solution for b_n   Continuing from the reuse point,

$$b_n \pi = (0, -\pi) + 2(0, \pi)$$

apply Rule 3b

$$= \frac{1}{n} [1 - \cos n(-\pi)] + \frac{2}{n}(1 - \cos n\pi)$$

$$= \frac{1}{n}(1 - \cos n\pi + 2 - 2 \cos n\pi)$$

$$b_n \pi = \frac{1}{n}(3 - 3 \cos n\pi)$$

$$b_n = \frac{3}{n\pi}(1 - \cos n\pi)$$

Therefore

$$b_n = \begin{cases} \dfrac{6}{n\pi} & n \text{ is odd} \\ 0 & n \text{ is even} \end{cases}$$

Solution for Fourier series

$$f(x) = \frac{a_0}{2} + \sum_{n=1}^{\infty} a_n \cos nx + \sum_{n=1}^{\infty} b_n \sin nx$$

$$= \frac{1}{2} + a_n \Big(\cos x + \cos 2x + \cdots \Big)$$

$$+ b_n \Big(\sin x + \sin 2x + \cdots \Big)$$

$$= \frac{1}{2} + 0 + b_n \Big(\sin x + \sin 3x + \cdots \Big)$$

$$+ b_n \Big(\sin 2x + \sin 4x + \cdots \Big)$$

$$= \frac{1}{2} + \frac{6}{n\pi} \Big(\sin x + \sin 3x + \cdots \Big)$$

$$+ 0 \Big(\sin 2x + \sin 4x + \cdots \Big)$$

$$\boxed{f(x) = \frac{1}{2} + \frac{6}{\pi}\left(\sin x + \frac{1}{3}\sin 3x + \cdots\right)}$$

3.

$$f(x) = \begin{cases} \dfrac{x}{2} & 0 < x < \pi \\[2mm] \pi - \dfrac{x}{2} & \pi < x < 2\pi \end{cases}$$

Solution for a_0

$a_0\pi$ = sum of cases

$$= \frac{x}{2}(0, \pi) + (\pi - \frac{x}{2})(\pi, 2\pi)$$

$$= \frac{1}{2}x(0, \pi) + \pi(\pi, 2\pi) - \frac{1}{2}x(\pi, 2\pi)$$

$$= \frac{1}{2}x(0, \pi) + \pi(\pi, 0) + \pi(0, 2\pi)$$

$$\quad - \frac{1}{2}x(\pi, 0) - \frac{1}{2}x(0, 2\pi)$$

$$= \frac{1}{2}x(0, \pi) - \pi(0, \pi) + \pi(0, 2\pi)$$

$$\quad + \frac{1}{2}x(0, \pi) - \frac{1}{2}x(0, 2\pi)$$

$$= x(0, \pi) - \pi(0, \pi) + \pi(0, 2\pi)$$

$$\quad - \frac{1}{2}x(0, 2\pi) \qquad \text{(RP)}$$

apply Rules 3c and 4c

$$= \frac{\pi^2}{2} - \pi(\pi) + \pi(2\pi) - \frac{1}{2}\cdot\frac{(2\pi)^2}{2}$$

$$= \frac{\pi^2}{2} - \pi^2 + 2\pi^2 - \frac{4\pi^2}{4}$$

$$a_0\pi = \frac{\pi^2}{2}$$

$$\boxed{a_0 = \frac{\pi}{2}}$$

Solution for a_n Continuing from the reuse point,

$$a_n\pi = x(0, \pi) - \pi(0, \pi) + \pi(0, 2\pi)$$

$$\quad - \frac{1}{2}x(0, 2\pi)$$

apply Rules 3a and 4a

$$= \left[\frac{\pi}{n}\sin n\pi + \frac{1}{n^2}(\cos n\pi - 1)\right]$$

$$\quad - \frac{\pi}{n}\sin n\pi + \frac{\pi}{n}\sin n(2\pi)$$

$$\quad - \frac{1}{2}\left[\frac{\pi}{n}\sin n(2\pi) + \frac{1}{n^2}(\cos n(2\pi) - 1)\right]$$

$$= \left[0 + \frac{1}{n^2}(\cos n\pi - 1)\right] - 0 + 0$$

$$\quad - \frac{1}{2}\left[0 + \frac{1}{n^2}(1 - 1)\right]$$

$$a_n\pi = \frac{1}{n^2}(\cos n\pi - 1)$$

$$a_n = \frac{1}{\pi n^2}(\cos n\pi - 1)$$

Therefore

$$a_n = \begin{cases} -\dfrac{2}{\pi n^2} & n \text{ is odd} \\[2mm] 0 & n \text{ is even} \end{cases}$$

Solution for b_n Continuing from the reuse point,

$$b_n\pi = x(0, \pi) - \pi(0, \pi) + \pi(0, 2\pi)$$

$$\quad - \frac{1}{2}x(0, 2\pi)$$

apply Rules 3b and 4b

$$= \left[-\frac{\pi}{n}\cos n\pi + \frac{1}{n^2}\sin n\pi\right]$$

$$\quad - \frac{\pi}{n}(1 - \cos n\pi) + \frac{\pi}{n}[1 - \cos n(2\pi)]$$

$$\quad - \frac{1}{2}\left[-\frac{2\pi}{n}\cos n(2\pi) + \frac{1}{n^2}\sin n(2\pi)\right]$$

$$= \left[-\frac{\pi}{n}\cos n\pi + 0\right]$$

$$\quad - \frac{\pi}{n}(1 - \cos n\pi) + \frac{\pi}{n}(1 - 1)$$

$$\quad - \frac{1}{2}\left[-\frac{2\pi}{n}(1) + 0\right]$$

$$= -\frac{\pi}{n}\cos n\pi - \frac{\pi}{n}(1 - \cos n\pi) + \frac{\pi}{n}$$

$$b_n\pi = \frac{\pi}{n}\left[-\cos n\pi - 1 + \cos n\pi + 1\right]$$

$$\boxed{b_n = 0}$$

Solution for Fourier series

$$f(x) = \frac{a_0}{2} + \sum_{n=1}^{\infty} a_n \cos nx + \sum_{n=1}^{\infty} b_n \sin nx$$

$$= \frac{\pi}{4} + a_n \Big(\cos x + \cos 2x + \cdots \Big)$$

$$+ b_n \Big(\sin x + \sin 2x + \cdots \Big)$$

$$= \frac{\pi}{4} + a_n \Big(\cos x + \cos 3x + \cdots \Big)$$

$$+ a_n \Big(\cos 2x + \cos 4x + \cdots \Big) + 0$$

$$= \frac{\pi}{4} - \frac{2}{\pi n^2} \Big(\cos x + \cos 3x + \cdots \Big)$$

$$+ 0 \Big(\cos 2x + \cos 4x + \cdots \Big)$$

$$\boxed{ f(x) = \frac{\pi}{4} - \frac{2}{\pi} \Big(\cos x + \frac{1}{3^2} \cos 3x + \cdots \Big) }$$

4.

$$f(x) = \begin{cases} -1 & -\pi < x < -\dfrac{\pi}{2} \\ 0 & -\dfrac{\pi}{2} < x < \dfrac{\pi}{2} \\ 1 & \dfrac{\pi}{2} < x < \pi \end{cases}$$

Solution for a_0

$a_0 \pi = $ sum of cases

$$= -1 \Big(-\pi, -\frac{\pi}{2} \Big) + 0 \Big(-\frac{\pi}{2}, \frac{\pi}{2} \Big) + 1 \Big(\frac{\pi}{2}, \pi \Big)$$

$$= -\Big(-\pi, -\frac{\pi}{2} \Big) + 0 + \Big(\frac{\pi}{2}, \pi \Big)$$

$$= -(-\pi, 0) - \Big(0, -\frac{\pi}{2} \Big) + \Big(\frac{\pi}{2}, 0 \Big) + (0, \pi)$$

$$= (0, -\pi) - \Big(0, -\frac{\pi}{2} \Big)$$

$$- \Big(0, \frac{\pi}{2} \Big) + (0, \pi) \qquad \text{(RP)}$$

apply Rule 3c

$$= (-\pi) - \Big(-\frac{\pi}{2} \Big) - \Big(\frac{\pi}{2} \Big) + (\pi)$$

$$a_0 \pi = -\pi + \frac{\pi}{2} - \frac{\pi}{2} + \pi$$

$$\boxed{ a_0 = 0 }$$

Solution for a_n Continuing from the reuse point,

$$a_n \pi = (0, -\pi) - \Big(0, -\frac{\pi}{2} \Big) - \Big(0, \frac{\pi}{2} \Big) + (0, \pi)$$

apply Rule 3a

$$= \frac{1}{n} \sin n(-\pi) - \frac{1}{n} \sin n \Big(-\frac{\pi}{2} \Big)$$

$$- \frac{1}{n} \sin n \Big(\frac{\pi}{2} \Big) + \frac{1}{n} \sin n(\pi)$$

$$a_n \pi = 0 + \frac{1}{n} \sin \frac{n\pi}{2} - \frac{1}{n} \sin \frac{n\pi}{2} + 0$$

$$\boxed{ a_n = 0 }$$

Solution for b_n Continuing from the reuse point,

$$b_n \pi$$

$$= (0, -\pi) - \Big(0, -\frac{\pi}{2} \Big) - \Big(0, \frac{\pi}{2} \Big) + (0, \pi)$$

apply Rule 3b

$$= \frac{1}{n} \Big[1 - \cos n(-\pi) \Big] - \frac{1}{n} \Big[1 - \cos n \Big(-\frac{\pi}{2} \Big) \Big]$$

$$- \frac{1}{n} \Big[1 - \cos n \Big(\frac{\pi}{2} \Big) \Big] + \frac{1}{n} \Big[1 - \cos n\pi \Big]$$

$$= \frac{1}{n} (1 - \cos n\pi) - \frac{1}{n} \Big(1 - \cos \frac{n\pi}{2} \Big)$$

$$- \frac{1}{n} \Big(1 - \cos \frac{n\pi}{2} \Big) + \frac{1}{n} (1 - \cos n\pi)$$

$$= \frac{2}{n} (1 - \cos n\pi) - \frac{2}{n} \Big(1 - \cos \frac{n\pi}{2} \Big)$$

$$= \frac{2}{n} \Big(1 - \cos n\pi - 1 + \cos \frac{n\pi}{2} \Big)$$

$$= \frac{2}{n} \Big(\cos \frac{n\pi}{2} - \cos n\pi \Big)$$

Thus

$$b_n = \frac{2}{n\pi} \Big(\cos \frac{n\pi}{2} - \cos n\pi \Big)$$

Recall that

$$\cos n\pi = \begin{cases} -1 & n \text{ is odd} \\ 1 & n \text{ is even} \end{cases}$$

and

$$\cos\frac{n\pi}{2} = \begin{cases} 0 & n \text{ is odd} \\ -1 & n = 2, 6, 10, \ldots \\ 1 & n = 4, 8, 12, \ldots \end{cases}$$

When n is odd

$$b_n = \frac{2}{n\pi}[0 - (-1)]$$

$$= \frac{2}{n\pi}$$

When $n = 2, 6, 10, \ldots$

$$b_n = \frac{2}{n\pi}(-1 - 1)$$

$$= -\frac{4}{n\pi}$$

When $n = 4, 8, 12, \ldots$

$$b_n = \frac{2}{n\pi}(1 - 1)$$

$$= 0$$

Therefore

$$b_n = \begin{cases} \dfrac{2}{n\pi} & n \text{ is odd} \\ -\dfrac{4}{n\pi} & n = 2, 6, 10, \ldots \\ 0 & n = 4, 8, 12, \ldots \end{cases}$$

Solution for Fourier series

$$f(x) = \frac{a_0}{2} + \sum_{n=1}^{\infty} a_n \cos nx + \sum_{n=1}^{\infty} b_n \sin nx$$

$$= \frac{0}{4} + 0\Big(\cos x + \cos 2x + \cdots\Big)$$

$$+ b_n\Big(\sin x + \sin 2x + \cdots\Big)$$

$$= 0 + 0 + b_n\Big(\sin x + \sin 2x + \cdots\Big)$$

$$= b_n\Big(\sin x + \sin 3x + \cdots\Big)$$

$$+ b_n\Big(\sin 2x + \sin 6x + \cdots\Big)$$

$$+ b_n\Big(\sin 4x + \sin 8x + \cdots\Big)$$

$$= \frac{2}{n\pi}\Big(\sin x + \sin 3x + \cdots\Big)$$

$$- \frac{4}{n\pi}\Big(\sin 2x + \sin 6x + \cdots\Big) + 0$$

$$\boxed{\begin{aligned} f(x) &= \frac{2}{\pi}\Big(\sin x + \frac{1}{3}\sin 3x + \cdots\Big) \\ &- \frac{4}{\pi}\Big(\frac{1}{2}\sin 2x + \frac{1}{6}\sin 6x + \cdots\Big) \end{aligned}}$$

5.

$$f(x) = \begin{cases} -x & -\pi < x < -\dfrac{\pi}{2} \\ 0 & -\dfrac{\pi}{2} < x < \dfrac{\pi}{2} \\ x & \dfrac{\pi}{2} < x < \pi \end{cases}$$

Solution for a_0

$$a_0\pi = \text{sum of cases}$$

$$= -x(-\pi, -\frac{\pi}{2}) + 0(-\frac{\pi}{2}, \frac{\pi}{2})$$

$$+ x(\frac{\pi}{2}, \pi)$$

$$= -x(-\pi, -\frac{\pi}{2}) + x(\frac{\pi}{2}, \pi)$$

$$= -x(-\pi, 0) - x(0, -\frac{\pi}{2})$$

$$+ x(\frac{\pi}{2}, 0) + x(0, \pi)$$

$$= x(0, -\pi) - x(0, -\frac{\pi}{2})$$

$$- x(0, \frac{\pi}{2}) + x(0, \pi) \qquad \text{(RP)}$$

apply Rule 4c

$$= \frac{(-\pi)^2}{2} - \frac{(-\pi/2)^2}{2} - \frac{(\pi/2)^2}{2} + \frac{\pi^2}{2}$$

$$a_0\pi = \frac{\pi^2}{2} - \frac{\pi^2}{8} - \frac{\pi^2}{8} + \frac{\pi^2}{2}$$

$$\boxed{a_0 = \frac{3\pi}{4}}$$

Solution for a_n Continuing from the reuse point,

$$a_n \pi$$

$$= x(0, -\pi) - x(0, -\frac{\pi}{2}) - x(0, \frac{\pi}{2}) + x(0, \pi)$$

apply Rule 4a

$$= \left[-\frac{\pi}{n} \sin n(-\pi) + \frac{1}{n^2}(\cos n(-\pi) - 1) \right]$$

$$- \left[\frac{-\pi/2}{n} \sin n\left(-\frac{\pi}{2} \right) \right.$$

$$\left. + \frac{1}{n^2}\left(\cos n\left(-\frac{\pi}{2} \right) - 1 \right) \right]$$

$$- \left[\frac{\pi/2}{n} \sin n\left(\frac{\pi}{2} \right) + \frac{1}{n^2}\left(\cos n\left(\frac{\pi}{2} \right) - 1 \right) \right]$$

$$+ \left[-\frac{\pi}{n} \sin n\pi + \frac{1}{n^2}(\cos n\pi - 1) \right]$$

$$= \left[-0 + \frac{1}{n^2}(\cos n\pi - 1) \right]$$

$$- \left[\frac{\pi}{2n} \sin \frac{n\pi}{2} + \frac{1}{n^2}\left(\cos \frac{n\pi}{2} - 1 \right) \right]$$

$$- \left[\frac{\pi}{2n} \sin \frac{n\pi}{2} + \frac{1}{n^2}\left(\cos \frac{n\pi}{2} - 1 \right) \right]$$

$$+ \left[-0 + \frac{1}{n^2}(\cos n\pi - 1) \right]$$

$$= \frac{2}{n^2}(\cos n\pi - 1)$$

$$- 2\left[\frac{\pi}{2n} \sin \frac{n\pi}{2} + \frac{1}{n^2}\left(\cos \frac{n\pi}{2} - 1 \right) \right]$$

$$= \frac{2}{n^2}(\cos n\pi - 1 - \cos \frac{n\pi}{2} + 1) - \frac{\pi}{n} \sin \frac{n\pi}{2}$$

Thus

$$a_n = \frac{2}{\pi n^2}(\cos n\pi - \cos \frac{n\pi}{2}) - \frac{1}{n} \sin \frac{n\pi}{2}$$

Recall that

$$\cos n\pi = \begin{cases} -1 & n \text{ is odd} \\ 1 & n \text{ is even} \end{cases}$$

$$\cos \frac{n\pi}{2} = \begin{cases} 0 & n \text{ is odd} \\ -1 & n = 2, 6, 10, \ldots \\ 1 & n = 4, 8, 12, \ldots \end{cases}$$

and

$$\sin \frac{n\pi}{2} = \begin{cases} 0 & n \text{ is even} \\ 1 & n = 1, 5, 19, \ldots \\ -1 & n = 3, 7, 11, \ldots \end{cases}$$

When $n = 1, 5, 19, \ldots$

$$a_n = \frac{2}{\pi n^2}(-1 - 0) - \frac{1}{n}(1)$$

$$= -\frac{2}{\pi n^2} - \frac{1}{n}$$

When $n = 2, 6, 10, \ldots$

$$a_n = \frac{2}{\pi n^2}[1 - (-1)] - \frac{1}{n}(0)$$

$$= \frac{4}{\pi n^2}$$

When $n = 3, 7, 11, \ldots$

$$a_n = \frac{2}{\pi n^2}(-1 - 0) - \frac{1}{n}(-1)$$

$$= -\frac{2}{\pi n^2} + \frac{1}{n}$$

When $n = 4, 8, 12, \ldots$

$$a_n = \frac{2}{\pi n^2}(1 - 1) - \frac{1}{n}(0)$$

$$= 0$$

Therefore

$$a_n = \begin{cases} -\dfrac{2}{\pi n^2} - \dfrac{1}{n} & n = 1, 5, 19, \ldots \\[2mm] \dfrac{4}{\pi n^2} & n = 2, 6, 10, \ldots \\[2mm] -\dfrac{2}{\pi n^2} + \dfrac{1}{n} & n = 3, 7, 11, \ldots \\[2mm] 0 & n = 4, 8, 12, \ldots \end{cases}$$

Solution for b_n Continuing from the reuse point,

$b_n\pi$

$= x(0, -\pi) - x(0, -\frac{\pi}{2}) - x(0, \frac{\pi}{2}) + x(0, \pi)$

apply Rule 4b

$= \left[-\frac{(-\pi)}{n} \cos n(-\pi) + \frac{1}{n^2} \sin n(-\pi) \right]$

$- \left[-\frac{(-\pi/2)}{n} \cos n(-\frac{\pi}{2}) + \frac{1}{n^2} \sin n(-\frac{\pi}{2}) \right]$

$- \left[-\frac{\pi/2}{n} \cos n(\frac{\pi}{2}) + \frac{1}{n^2} \sin n(\frac{\pi}{2}) \right]$

$+ \left[-\frac{\pi}{n} \cos n\pi + \frac{1}{n^2} \sin n\pi \right]$

$= \left[\frac{\pi}{n} \cos n\pi + 0 \right]$

$- \left[\frac{\pi}{2n} \cos \frac{n\pi}{2} - \frac{1}{n^2} \sin \frac{n\pi}{2} \right]$

$- \left[-\frac{\pi}{2n} \cos \frac{n\pi}{2} + \frac{1}{n^2} \sin \frac{n\pi}{2} \right]$

$+ \left[-\frac{\pi}{n} \cos n\pi + 0 \right]$

$= \frac{\pi}{n} \cos n\pi - \frac{\pi}{n} \cos n\pi$

$- \frac{\pi}{2n} \cos \frac{n\pi}{2} + \frac{\pi}{2n} \cos \frac{n\pi}{2}$

$- \frac{1}{n^2} \sin \frac{n\pi}{2} + \frac{1}{n^2} \sin \frac{n\pi}{2}$

$= 0$

Thus

$$\boxed{b_n = 0}$$

Solution for Fourier series

$f(x) = \frac{a_0}{2} + \sum_{n=1}^{\infty} a_n \cos nx + \sum_{n=1}^{\infty} b_n \sin nx$

$= \frac{3\pi}{8} + a_n \left(\cos x + \cos 2x + \cdots \right)$

$+ 0 \left(\sin x + \sin 2x + \cdots \right)$

$= \frac{3\pi}{8} + a_n \left(\cos x + \cos 5x + \cdots \right)$

$+ a_n \left(\cos 2x + \cos 6x + \cdots \right)$

$+ a_n \left(\cos 3x + \cos 7x + \cdots \right)$

$+ a_n \left(\cos 4x + \cos 8x + \cdots \right)$

$= \frac{3\pi}{8}$

$+ \left(-\frac{2}{\pi n^2} - \frac{1}{n} \right) \left(\cos x + \cos 5x + \cdots \right)$

$+ \frac{4}{\pi n^2} \left(\cos 2x + \cos 6x + \cdots \right)$

$+ \left(-\frac{2}{\pi n^2} + \frac{1}{n} \right) \left(\cos 3x + \cos 7x + \cdots \right)$

$+ 0 \left(\cos 4x + \cos 8x + \cdots \right)$

$= \frac{3\pi}{8} - \frac{2}{\pi n^2} \left(\cos x + \cos 5x + \cdots \right)$

$- \frac{1}{n} \left(\cos x + \cos 5x + \cdots \right)$

$+ \frac{4}{\pi n^2} \left(\cos 2x + \cos 6x + \cdots \right)$

$- \frac{2}{\pi n^2} \left(\cos 3x + \cos 7x + \cdots \right)$

$+ \frac{1}{n} \left(\cos 3x + \cos 7x + \cdots \right)$

$= \frac{3\pi}{8} - \frac{2}{\pi n^2} \left(\cos x + \cos 3x + \cdots \right)$

$- \frac{1}{n} \left(\cos x - \cos 3x + \cdots \right)$

$+ \frac{4}{\pi n^2} \left(\cos 2x + \cos 6x + \cdots \right)$

$$\boxed{f(x) = \frac{3\pi}{8} - \frac{2}{\pi} \left(\cos x + \frac{1}{3^2} \cos 3x + \cdots \right) \\ - \left(\cos x - \frac{1}{3} \cos 3x + \cdots \right) \\ + \frac{4}{\pi} \left(\frac{1}{2^2} \cos 2x + \frac{1}{6^2} \cos 6x + \cdots \right)}$$

Exercise 2

1.

$f(x) = |x| \qquad -\pi < x < \pi$

Recall that

$|x| = \begin{cases} -x & x < 0 \\ x & x \geq 0 \end{cases}$

Therefore

$$f(x) = \begin{cases} -x & -\pi < x < 0 \\ x & 0 \le x < \pi \end{cases}$$

Solution for Fourier series

$$f(x) = \frac{a_0}{2} + \sum_{n=1}^{\infty} a_n \cos nx + \sum_{n=1}^{\infty} b_n \sin nx$$

$$= \frac{\pi}{2} + a_n \left(\cos x + \cos 3x + \cdots \right)$$

$$+ a_n \left(\cos 2x + \cos 4x + \cdots \right)$$

$$+ b_n \left(\sin x + \sin 2x + \cdots \right)$$

$$= \frac{\pi}{2} - \frac{4}{\pi n^2} \left(\cos x + \cos 3x + \cdots \right)$$

$$+ 0 \left(\cos 2x + \cos 4x + \cdots \right)$$

$$+ 0 \left(\sin x + \sin 2x + \cdots \right)$$

$$\boxed{f(x) = \frac{\pi}{2} - \frac{4}{\pi} \left(\cos x + \frac{1}{3^2} \cos 3x + \cdots \right)}$$

Solution for b_n

$b_n \pi =$ sum of cases

$$= -x(-\pi, 0) + x(0, \pi)$$

$$= x(0, -\pi) + x(0, \pi) \qquad \text{(RP1)}$$

apply Rule 6b

$$b_n \pi = -x(0, \pi) + x(0, \pi)$$

$$\boxed{b_n = 0}$$

Solution for a_n   Continuing from RP1,

$$a_n \pi = x(0, -\pi) + x(0, \pi)$$

apply Rule 6a

$$= x(0, \pi) + x(0, \pi)$$

$$= 2x(0, \pi) \qquad \text{(RP2)}$$

$$= 2 \left[\frac{\pi}{n} \sin n\pi + \frac{1}{n^2} (\cos n\pi - 1) \right]$$

$$a_n \pi = 2 \left[0 + \frac{1}{n^2} (\cos n\pi - 1) \right]$$

$$a_n = \frac{2}{\pi n^2} (\cos n\pi - 1)$$

Therefore

$$\boxed{a_n = \begin{cases} -\dfrac{4}{\pi n^2} & n \text{ is odd} \\ 0 & n \text{ is even} \end{cases}}$$

2.

$$f(x) = \pi - |x| \qquad -\pi < x < \pi$$

Recall that from the previous problem (1) above,

$$|x| = \frac{\pi}{2} - \frac{4}{\pi} \left(\cos x + \frac{1}{3^2} \cos 3x + \cdots \right)$$

$$-|x| = -\frac{\pi}{2} + \frac{4}{\pi} \left(\cos x + \frac{1}{3^2} \cos 3x + \cdots \right)$$

$$\pi - |x| = \pi - \frac{\pi}{2} + \frac{4}{\pi} \left(\cos x + \frac{1}{3^2} \cos 3x + \cdots \right)$$

$$\boxed{f(x) = \frac{\pi}{2} + \frac{4}{\pi} \left(\cos x + \frac{1}{3^2} \cos 3x + \cdots \right)}$$

Solution for a_0   Continuing from RP2,

$$a_0 \pi = 2x(0, \pi)$$

$$= 2 \times \frac{\pi^2}{2}$$

$$= \pi^2$$

$$\boxed{a_0 = \pi}$$

3.

$$f(x) = \begin{cases} x + \pi & 0 < x < \pi \\ -x - \pi & -\pi < x < 0 \end{cases}$$

Solution for b_n

$b_n \pi =$ sum of cases

$= (x + \pi)(0, \pi) + (-x - \pi)(-\pi, 0)$

$= x(0, \pi) + \pi(0, \pi) - x(-\pi, 0) - \pi(-\pi, 0)$

$= x(0, \pi) + \pi(0, \pi)$

$\qquad + x(0, -\pi) + \pi(0, -\pi)$ (RP1)

apply Rules 5b and 6b

$= x(0, \pi) + \pi(0, \pi) - x(0, \pi) + \pi(0, \pi)$

$= 2\pi(0, \pi)$

$b_n \pi = 2 \cdot \dfrac{\pi}{n}(1 - \cos n\pi)$

$b_n = \dfrac{2}{n}(1 - \cos n\pi)$

Therefore

$$b_n = \begin{cases} \dfrac{4}{n} & n \text{ is odd} \\[2mm] 0 & n \text{ is even} \end{cases}$$

Solution for a_n   Continuing from RP1,

$a_n \pi = x(0, \pi) + \pi(0, \pi) + x(0, -\pi) + \pi(0, -\pi)$

apply Rules 5a and 6a

$= x(0, \pi) + \pi(0, \pi) + x(0, \pi) - \pi(0, \pi)$

$= 2x(0, \pi)$ (RP2)

$a_n \pi = 2\left[\dfrac{\pi}{n} \sin n\pi + \dfrac{1}{n^2}(\cos n\pi - 1) \right]$

$= 2\left[0 + \dfrac{1}{n^2}(\cos n\pi - 1) \right]$

$a_n \pi = \dfrac{2}{n^2}(\cos n\pi - 1)$

$a_n = \dfrac{2}{\pi n^2}(\cos n\pi - 1)$

Therefore

$$a_n = \begin{cases} -\dfrac{4}{\pi n^2} & n \text{ is odd} \\[2mm] 0 & n \text{ is even} \end{cases}$$

Solution for a_0   Continuing from RP2,

$a_0 \pi = 2x(0, \pi)$

$= 2 \times \dfrac{\pi^2}{2}$

$= \pi^2$

$\boxed{a_0 = \pi}$

Solution for Fourier series

$f(x) = \dfrac{a_0}{2} + \displaystyle\sum_{n=1}^{\infty} a_n \cos nx + \sum_{n=1}^{\infty} b_n \sin nx$

$= \dfrac{\pi}{2} + a_n\left(\cos x + \cos 3x + \cdots \right)$

$\qquad + a_n\left(\cos 2x + \cos 4x + \cdots \right)$

$\qquad + b_n\left(\sin x + \sin 3x + \cdots \right)$

$\qquad + b_n\left(\sin 2x + \sin 4x + \cdots \right)$

$= \dfrac{\pi}{2} - \dfrac{4}{\pi n^2}\left(\cos x + \cos 3x + \cdots \right)$

$\qquad + 0\left(\cos 2x + \cos 4x + \cdots \right)$

$\qquad + \dfrac{4}{n}\left(\sin x + \sin 3x + \cdots \right)$

$\qquad + 0\left(\sin 2x + \sin 4x + \cdots \right)$

$$\boxed{\begin{aligned} f(x) = {} & \dfrac{\pi}{2} - \dfrac{4}{\pi}\left(\cos x + \dfrac{1}{3^2} \cos 3x + \cdots \right) \\ & + 4\left(\sin x + \dfrac{1}{3} \sin 3x + \cdots \right) \end{aligned}}$$

4.

$$f(x) = \begin{cases} -\pi & -\pi < x < 0 \\ x & 0 < x < \pi \end{cases}$$

Solution for b_n

$b_n \pi =$ sum of cases

$= -\pi(-\pi, 0) + x(0, \pi)$

$= \pi(0, -\pi) + x(0, \pi)$ (RP1)

apply Rule 5b

$= \pi(0, \pi) + x(0, \pi)$

$= \dfrac{\pi}{n}(1 - \cos n\pi)$

$\qquad + \left[-\dfrac{\pi}{n} \cos n\pi + \dfrac{1}{n^2} \sin n\pi \right]$

$$= \frac{\pi}{n}(1 - \cos n\pi) + \left[-\frac{\pi}{n}\cos n\pi + 0 \right]$$

$$b_n\pi = \frac{\pi}{n}(1 - \cos n\pi - \cos n\pi)$$

$$b_n = \frac{1}{n}(1 - 2\cos n\pi)$$

Therefore

$$b_n = \begin{cases} \dfrac{3}{n} & n \text{ is odd} \\ -\dfrac{1}{n} & n \text{ is even} \end{cases}$$

Solution for a_n   Continuing from RP1,

$$a_n\pi = \pi(0, -\pi) + x(0, \pi)$$

apply Rule 5a

$$= -\pi(0, \pi) + x(0, \pi) \qquad \text{(RP2)}$$

$$= -\frac{\pi}{n}\sin n\pi$$

$$+ \left[\frac{\pi}{n}\sin n\pi + \frac{1}{n^2}(\cos n\pi - 1) \right]$$

$$= 0 + \left[0 + \frac{1}{n^2}(\cos n\pi - 1) \right]$$

$$a_n\pi = \frac{1}{n^2}(\cos n\pi - 1)$$

$$a_n = \frac{1}{\pi n^2}(\cos n\pi - 1)$$

Therefore

$$a_n = \begin{cases} -\dfrac{2}{\pi n^2} & n \text{ is odd} \\ 0 & n \text{ is even} \end{cases}$$

Solution for a_0   Continuing from RP2,

$$a_0\pi = -\pi(0, \pi) + x(0, \pi)$$

$$= -\pi(\pi) + \frac{\pi^2}{2}$$

$$a_0\pi = -\frac{\pi^2}{2}$$

$$a_0 = -\frac{\pi}{2}$$

Solution for Fourier series

$$f(x) = \frac{a_0}{2} + \sum_{n=1}^{\infty} a_n \cos nx + \sum_{n=1}^{\infty} b_n \sin nx$$

$$= -\frac{\pi}{4} + a_n\left(\cos x + \cos 3x + \cdots \right)$$

$$+ a_n\left(\cos 2x + \cos 4x + \cdots \right)$$

$$+ b_n\left(\sin x + \sin 3x + \cdots \right)$$

$$+ b_n\left(\sin 2x + \sin 4x + \cdots \right)$$

$$= -\frac{\pi}{4} - \frac{2}{\pi n^2}\left(\cos x + \cos 3x + \cdots \right)$$

$$+ 0\left(\cos 2x + \cos 4x + \cdots \right)$$

$$+ \frac{3}{n}\left(\sin x + \sin 3x + \cdots \right)$$

$$- \frac{1}{n}\left(\sin 2x + \sin 4x + \cdots \right)$$

$$f(x) = -\frac{\pi}{4} - \frac{2}{\pi}\left(\cos x + \frac{1}{3^2}\cos 3x + \cdots \right)$$

$$+ 3\left(\sin x + \frac{1}{3}\sin 3x + \cdots \right)$$

$$- \left(\frac{1}{2}\sin 2x + \frac{1}{4}\sin 4x + \cdots \right)$$

5.

$$f(x) = \begin{cases} 0 & -\pi < x < 0 \\ 1 & 0 < x < \dfrac{\pi}{2} \\ 0 & \dfrac{\pi}{2} < x < \pi \end{cases}$$

Solution for b_n

$$b_n\pi = \text{sum of cases}$$

$$= 0(-\pi, 0) + 1(0, \frac{\pi}{2}) + 0(\frac{\pi}{2}, \pi)$$

$$= (0, \frac{\pi}{2}) \qquad \text{(RP1)}$$

$$b_n\pi = \frac{1}{n}(1 - \cos\frac{n\pi}{2})$$

$$b_n = \frac{1}{n\pi}(1 - \cos\frac{n\pi}{2})$$

Recall that

$$\cos \frac{n\pi}{2} = \begin{cases} 0 & n \text{ is odd} \\ -1 & n = 2, 6, 10, \ldots \\ 1 & n = 4, 8, 12, \ldots \end{cases}$$

Therefore

$$b_n = \begin{cases} \dfrac{1}{n\pi} & n \text{ is odd} \\ \dfrac{2}{n\pi} & n = 2, 6, 10, \ldots \\ 0 & n = 4, 8, 12, \ldots \end{cases}$$

Solution for a_n   Continuing from RP1,

$$a_n \pi = (0, \frac{\pi}{2})$$ (RP2)

$$= \frac{1}{n} \sin \frac{n\pi}{2}$$

$$a_n = \frac{1}{n\pi} \sin \frac{n\pi}{2}$$

Recall that

$$\sin \frac{n\pi}{2} = \begin{cases} 0 & n \text{ is even} \\ 1 & n = 1, 5, 9, \ldots \\ -1 & n = 3, 7, 11, \ldots \end{cases}$$

Therefore

$$a_n = \begin{cases} 0 & n \text{ is even} \\ \dfrac{1}{n\pi} & n = 1, 5, 9, \ldots \\ -\dfrac{1}{n\pi} & n = 3, 7, 11, \ldots \end{cases}$$

Solution for a_0   Continuing from RP2,

$$a_0 \pi = (0, \frac{\pi}{2})$$

$$= \frac{\pi}{2}$$

$$a_0 = \frac{1}{2}$$

Solution for Fourier series

$$f(x) = \frac{a_0}{2} + \sum_{n=1}^{\infty} a_n \cos nx + \sum_{n=1}^{\infty} b_n \sin nx$$

$$= \frac{1}{4} + a_n \Big(\cos x + \cos 5x + \cdots \Big)$$

$$+ a_n \Big(\cos 3x + \cos 7x + \cdots \Big)$$

$$+ b_n \Big(\sin x + \sin 3x + \cdots \Big)$$

$$+ b_n \Big(\sin 2x + \sin 6x + \cdots \Big)$$

$$= \frac{1}{4} + \frac{1}{n\pi} \Big(\cos x + \cos 3x + \cdots \Big)$$

$$- \frac{1}{n\pi} \Big(\cos 2x + \cos 4x + \cdots \Big)$$

$$+ \frac{1}{n\pi} \Big(\sin x + \sin 3x + \cdots \Big)$$

$$+ \frac{2}{n\pi} \Big(\sin 2x + \sin 6x + \cdots \Big)$$

$$= \frac{1}{4} + \frac{1}{\pi} \Big(\cos x + \frac{1}{5} \cos 5x + \cdots \Big)$$

$$- \frac{1}{\pi} \Big(\frac{1}{3} \cos 2x + \frac{1}{7} \cos 7x + \cdots \Big)$$

$$+ \frac{1}{\pi} \Big(\sin x + \frac{1}{3} \sin 3x + \cdots \Big)$$

$$+ \frac{2}{\pi} \Big(\frac{1}{2} \sin 2x + \frac{1}{6} \sin 6x + \cdots \Big)$$

$$f(x) = \frac{1}{4} + \frac{1}{\pi} \Big(\cos x - \frac{1}{3} \cos 3x + \cdots \Big)$$

$$+ \frac{1}{\pi} \Big(\sin x + \frac{1}{3} \sin 3x + \cdots \Big)$$

$$+ \frac{2}{\pi} \Big(\frac{1}{2} \sin 2x + \frac{1}{6} \sin 6x + \cdots \Big)$$

6.

$$f(x) = \begin{cases} \pi + x & -\pi < x < -\dfrac{\pi}{2} \\ \dfrac{\pi}{2} & -\dfrac{\pi}{2} < x < \dfrac{\pi}{2} \\ \pi - x & \dfrac{\pi}{2} < x < \pi \end{cases}$$

Solution for b_n

$b_n \pi$ = sum of cases

$$= (\pi + x)(-\pi, -\frac{\pi}{2}) + \frac{\pi}{2}(-\frac{\pi}{2}, \frac{\pi}{2})$$

$$+ (\pi - x)(\frac{\pi}{2}, \pi)$$

$$= \pi(-\pi, -\frac{\pi}{2}) + x(-\pi, -\frac{\pi}{2}) + \frac{\pi}{2}(-\frac{\pi}{2}, \frac{\pi}{2})$$

$$+ \pi(\frac{\pi}{2}, \pi) - x(\frac{\pi}{2}, \pi)$$

$$= \pi(-\pi, -\frac{\pi}{2}) - \pi(\pi, \frac{\pi}{2})$$

$$+ x(-\pi, -\frac{\pi}{2}) + x(\pi, \frac{\pi}{2})$$

$$+ \frac{\pi}{2}(-\frac{\pi}{2}, 0) + \frac{\pi}{2}(0, \frac{\pi}{2}) \qquad \text{(RP1)}$$

apply Rules 5b and 6b

$$= \pi(\pi, \frac{\pi}{2}) - \pi(\pi, \frac{\pi}{2})$$

$$- x(\pi, \frac{\pi}{2}) + x(\pi, \frac{\pi}{2})$$

$$+ \frac{\pi}{2}(\frac{\pi}{2}, 0) + \frac{\pi}{2}(0, \frac{\pi}{2})$$

$$b_n \pi = -\frac{\pi}{2}(0, \frac{\pi}{2}) + \frac{\pi}{2}(0, \frac{\pi}{2})$$

$$\boxed{b_n = 0}$$

Solution for a_n   Continuing from RP1,

$$a_n \pi = \pi(-\pi, -\frac{\pi}{2}) - \pi(\pi, \frac{\pi}{2})$$

$$+ x(-\pi, -\frac{\pi}{2}) + x(\pi, \frac{\pi}{2})$$

$$+ \frac{\pi}{2}(-\frac{\pi}{2}, 0) + \frac{\pi}{2}(0, \frac{\pi}{2})$$

apply Rules 5a and 6a

$$= -\pi(\pi, \frac{\pi}{2}) - \pi(\pi, \frac{\pi}{2})$$

$$+ x(\pi, \frac{\pi}{2}) + x(\pi, \frac{\pi}{2})$$

$$- \frac{\pi}{2}(\frac{\pi}{2}, 0) + \frac{\pi}{2}(0, \frac{\pi}{2})$$

$$= -2\pi(\pi, \frac{\pi}{2}) + 2x(\pi, \frac{\pi}{2})$$

$$+ \frac{\pi}{2}(0, \frac{\pi}{2}) + \frac{\pi}{2}(0, \frac{\pi}{2})$$

$$= -2\pi(\pi, 0) - 2\pi(0, \frac{\pi}{2})$$

$$+ 2x(\pi, 0) + 2x(0, \frac{\pi}{2}) + \pi(0, \frac{\pi}{2})$$

$$= 2\pi(0, \pi) - \pi(0, \frac{\pi}{2})$$

$$- 2x(0, \pi) + 2x(0, \frac{\pi}{2}) \qquad \text{(RP2)}$$

$$= 2\frac{\pi}{n} \sin n\pi - \frac{\pi}{n} \sin \frac{n\pi}{2}$$

$$- 2\left[\frac{\pi}{n} \sin n\pi + \frac{1}{n^2}(\cos n\pi - 1)\right]$$

$$+ 2\left[\frac{\pi/2}{n} \sin \frac{n\pi}{2} + \frac{1}{n^2}(\cos \frac{n\pi}{2} - 1)\right]$$

$$= 0 - \frac{\pi}{n} \sin \frac{n\pi}{2} - 2\left[0 + \frac{1}{n^2}(\cos n\pi - 1)\right]$$

$$+ \left[\frac{\pi}{n} \sin \frac{n\pi}{2} + \frac{2}{n^2}(\cos \frac{n\pi}{2} - 1)\right]$$

$$= -\frac{2}{n^2}(\cos n\pi - 1)$$

$$+ \frac{2}{n^2}(\cos \frac{n\pi}{2} - 1)$$

$$a_n \pi = \frac{2}{n^2}(-\cos n\pi + 1 + \cos \frac{n\pi}{2} - 1)$$

$$a_n = \frac{2}{\pi n^2}(\cos \frac{n\pi}{2} - \cos n\pi)$$

Recall that

$$\cos n\pi = \begin{cases} -1 & n \text{ is odd} \\ 1 & n \text{ is even} \end{cases}$$

and

$$\cos \frac{n\pi}{2} = \begin{cases} 0 & n \text{ is odd} \\ -1 & n = 2, 6, 10, \ldots \\ 1 & n = 4, 8, 12, \ldots \end{cases}$$

When n is odd

$$a_n = \frac{2}{\pi n^2}[0 - (-1)]$$

$$= \frac{2}{\pi n^2}$$

When $n = 2, 6, 10, \ldots$

$$a_n = \frac{2}{\pi n^2}(-1 - 1)$$

$$= -\frac{4}{\pi n^2}$$

When $n = 4, 8, 12, \ldots$

$$a_n = \frac{2}{\pi n^2}(1 - 1)$$

$$= 0$$

Therefore

$$a_n = \begin{cases} \dfrac{2}{\pi n^2} & n \text{ is odd} \\[2mm] -\dfrac{4}{\pi n^2} & n = 2, 6, 10, \ldots \\[2mm] 0 & n = 4, 8, 12, \ldots \end{cases}$$

Solution for a_0 Continuing from RP2,

$$a_0\pi = 2\pi(0, \pi) - \pi(0, \frac{\pi}{2}) - 2x(0, \pi) + 2x(0, \frac{\pi}{2})$$

$$= 2\pi(\pi) - \pi(\frac{\pi}{2}) - 2 \cdot \frac{\pi^2}{2} + 2 \cdot \frac{(\pi/2)^2}{2}$$

$$= 2\pi^2 - \frac{\pi^2}{2} - \pi^2 + \frac{\pi^2}{4}$$

$$a_0\pi = \frac{3\pi^2}{4}$$

$$\boxed{a_0 = \frac{3\pi}{4}}$$

Solution for Fourier series

$$f(x) = \frac{a_0}{2} + \sum_{n=1}^{\infty} a_n \cos nx + \sum_{n=1}^{\infty} b_n \sin nx$$

$$= \frac{3\pi}{8} + a_n\left(\cos x + \cos 3x + \cdots\right)$$

$$+ a_n\left(\cos 2x + \cos 6x + \cdots\right)$$

$$+ a_n\left(\cos 4x + \cos 8x + \cdots\right)$$

$$+ 0\left(\sin x + \sin 2x + \cdots\right)$$

$$= \frac{3\pi}{8} + \frac{2}{\pi n^2}\left(\cos x + \cos 3x + \cdots\right)$$

$$- \frac{4}{\pi n^2}\left(\cos 2x + \cos 6x + \cdots\right)$$

$$+ 0\left(\cos 4x + \cos 8x + \cdots\right)$$

$$\boxed{f(x) = \frac{3\pi}{8} + \frac{2}{\pi}\left(\cos x + \frac{1}{3^2}\cos 3x + \cdots\right) - \frac{4}{\pi}\left(\frac{1}{2^2}\cos 2x + \frac{1}{6^2}\cos 6x + \cdots\right)}$$

Exercise 3

1.

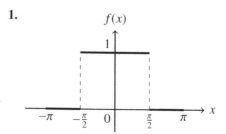

Solution for analytical definition

$$f(x) = \begin{cases} 0 & -\pi < x < -\dfrac{\pi}{2} \\[2mm] 1 & -\dfrac{\pi}{2} < x < \dfrac{\pi}{2} \\[2mm] 0 & \dfrac{\pi}{2} < x < \pi \end{cases}$$

Solution for b_n

$$b_n\pi = \text{sum of cases}$$

$$= 0(-\pi, -\frac{\pi}{2}) + 1(-\frac{\pi}{2}, \frac{\pi}{2}) + 0(\frac{\pi}{2}, \pi)$$

$$= (-\frac{\pi}{2}, \frac{\pi}{2})$$

$$= (-\frac{\pi}{2}, 0) + (0, \frac{\pi}{2})$$

$$= -(0, -\frac{\pi}{2}) + (0, \frac{\pi}{2}) \qquad \text{(RP1)}$$

apply Rule 5b

$$b_n\pi = -(0, \frac{\pi}{2}) + (0, \frac{\pi}{2})$$

$$\boxed{b_n = 0}$$

Solution for a_n Continuing from RP1,

$$a_n \pi = -(0, -\frac{\pi}{2}) + (0, \frac{\pi}{2})$$

apply Rule 5a

$$= (0, \frac{\pi}{2}) + (0, \frac{\pi}{2})$$

$$= 2(0, \frac{\pi}{2}) \qquad \text{(RP2)}$$

$$a_n \pi = 2\frac{1}{n} \sin \frac{n\pi}{2}$$

$$a_n = \frac{2}{n\pi} \sin \frac{n\pi}{2}$$

Recall that

$$\sin \frac{n\pi}{2} = \begin{cases} 0 & n \text{ is even} \\ 1 & n = 1, 5, 9, \ldots \\ -1 & n = 3, 7, 11, \ldots \end{cases}$$

Therefore

$$a_n = \begin{cases} 0 & n \text{ is even} \\ \dfrac{2}{n\pi} & n = 1, 5, 9, \ldots \\ -\dfrac{2}{n\pi} & n = 3, 7, 11, \ldots \end{cases}$$

Solution for a_0 Continuing from RP2,

$$a_0 \pi = 2(0, \frac{\pi}{2})$$

$$= 2(\frac{\pi}{2})$$

$$= \pi$$

$$\boxed{a_0 = 1}$$

Solution for Fourier series

$$f(x) = \frac{a_0}{2} + \sum_{n=1}^{\infty} a_n \cos nx + \sum_{n=1}^{\infty} b_n \sin nx$$

$$= \frac{1}{2} + a_n\left(\cos 2x + \cos 4x + \cdots\right)$$

$$+ a_n\left(\cos x + \cos 5x + \cdots\right)$$

$$+ a_n\left(\cos 3x + \cos 7x + \cdots\right)$$

$$+ 0\left(\sin x + \sin 2x + \cdots\right)$$

$$= \frac{1}{2} + 0\left(\cos 2x + \cos 4x + \cdots\right)$$

$$+ \frac{2}{n\pi}\left(\cos x + \cos 5x + \cdots\right)$$

$$- \frac{2}{n\pi}\left(\cos 3x + \cos 7x + \cdots\right)$$

$$= \frac{1}{2} + \frac{2}{\pi}\left(\cos x + \frac{1}{5}\cos 5x + \cdots\right)$$

$$- \frac{2}{\pi}\left(\frac{1}{3}\cos 3x + \frac{1}{7}\cos 7x + \cdots\right)$$

$$\boxed{f(x) = \frac{1}{2} + \frac{2}{\pi}\left(\cos x - \frac{1}{3}\cos 3x + \cdots\right)}$$

2.

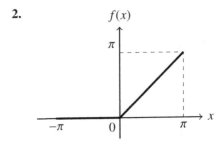

$$f(x)$$

Solution for analytical definition

$$f(x) = \begin{cases} 0 & -\pi < x < 0 \\ x & 0 < x < \pi \end{cases}$$

Solution for b_n

$$b_n \pi = \text{sum of cases}$$

$$= 0(-\pi, 0) + x(0, \pi)$$

$$= x(0, \pi) \qquad \text{(RP1)}$$

$$= -\frac{\pi}{n} \cos n\pi + \frac{1}{n^2} \sin n\pi$$

$$b_n \pi = -\frac{\pi}{n} \cos n\pi + 0$$

$$b_n = -\frac{1}{n} \cos n\pi$$

Therefore

$$b_n = \begin{cases} \dfrac{1}{n} & n \text{ is odd} \\ -\dfrac{1}{n} & n \text{ is even} \end{cases}$$

Solution for a_n   Continuing from RP1,

$$a_n\pi = x(0,\pi) \qquad \text{(RP2)}$$

$$= \frac{\pi}{n}\sin n\pi + \frac{1}{n^2}(\cos n\pi - 1)$$

$$= \frac{1}{n^2}(\cos n\pi - 1)$$

$$a_n = \frac{1}{\pi n^2}(\cos n\pi - 1)$$

Therefore

$$a_n = \begin{cases} -\dfrac{2}{\pi n^2} & n \text{ is odd} \\ 0 & n \text{ is even} \end{cases}$$

Solution for a_0   Continuing from RP2,

$$a_0\pi = x(0,\pi)$$

$$= \frac{\pi^2}{2}$$

$$a_0 = \frac{\pi}{2}$$

Solution for Fourier series

$$f(x) = \frac{a_0}{2} + \sum_{n=1}^{\infty} a_n \cos nx + \sum_{n=1}^{\infty} b_n \sin nx$$

$$= \frac{\pi}{4} + a_n\left(\cos x + \cos 3x + \cdots\right)$$

$$+ a_n\left(\cos 2x + \cos 4x + \cdots\right)$$

$$+ b_n\left(\sin x + \sin 3x + \cdots\right)$$

$$+ b_n\left(\sin 2x + \sin 4x + \cdots\right)$$

$$= \frac{\pi}{4} - \frac{2}{\pi n^2}\left(\cos x + \cos 3x + \cdots\right)$$

$$+ 0\left(\cos 2x + \cos 4x + \cdots\right)$$

$$+ \frac{1}{n}\left(\sin x + \sin 3x + \cdots\right)$$

$$- \frac{1}{n}\left(\sin 2x + \sin 4x + \cdots\right)$$

$$= \frac{\pi}{4} - \frac{2}{\pi}\left(\cos x + \frac{1}{3^2}\cos 3x + \cdots\right)$$

$$+ \left(\sin x + \frac{1}{3}\sin 3x + \cdots\right)$$

$$- \left(\frac{1}{2}\sin 2x + \frac{1}{4}\sin 4x + \cdots\right)$$

$$\boxed{\begin{aligned} f(x) = &\frac{\pi}{4} - \frac{2}{\pi}\left(\cos x + \frac{1}{3^2}\cos 3x + \cdots\right) \\ &+ \left(\sin x - \frac{1}{2}\sin 2x + \cdots\right) \end{aligned}}$$

3.

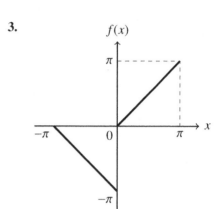

$f(x)$

Solution for analytical definition

$$f(x) = \begin{cases} -x - \pi & -\pi < x < 0 \\ x & 0 < x < \pi \end{cases}$$

Solution for b_n

$$b_n\pi = \text{sum of cases}$$

$$= (-x-\pi)(-\pi,0) + x(0,\pi)$$

$$= -x(-\pi,0) - \pi(-\pi,0) + x(0,\pi)$$

$$= x(0,-\pi) + \pi(0,-\pi)$$

$$\qquad + x(0,\pi) \qquad \text{(RP1)}$$

apply Rules 5b and 6b

$$= -x(0,\pi) + \pi(0,\pi) + x(0,\pi)$$

$$= \pi(0,\pi)$$

$$b_n\pi = \pi \cdot \frac{1}{n}(1 - \cos n\pi)$$

$$b_n = \frac{1}{n}(1 - \cos n\pi)$$

Therefore

$$b_n = \begin{cases} \dfrac{2}{n} & n \text{ is odd} \\ 0 & n \text{ is even} \end{cases}$$

$$f(x) = -\frac{4}{\pi}\left(\cos x + \frac{1}{3^2}\cos 3x + \cdots\right)$$
$$+ 2\left(\sin x + \frac{1}{3}\sin 3x + \cdots\right)$$

Solution for a_n   Continuing from RP1,

$$a_n\pi = x(0, -\pi) + \pi(0, -\pi) + x(0, \pi)$$

apply Rules 5a and 6a

$$= x(0, \pi) - \pi(0, \pi) + x(0, \pi)$$
$$= 2x(0, \pi) - \pi(0, \pi) \qquad \text{(RP2)}$$
$$= 2\left[\frac{\pi}{n}\sin n\pi + \frac{1}{n^2}(\cos n\pi - 1)\right]$$
$$- \frac{\pi}{n}\sin n\pi$$

$$a_n\pi = \frac{2}{n^2}(\cos n\pi - 1)$$

$$a_n = \frac{2}{\pi n^2}(\cos n\pi - 1)$$

Therefore

$$a_n = \begin{cases} -\dfrac{4}{\pi n^2} & n \text{ is odd} \\ 0 & n \text{ is even} \end{cases}$$

Solution for a_0   Continuing from RP2,

$$a_0\pi = 2x(0, \pi) - \pi(0, \pi)$$
$$= 2 \cdot \frac{\pi^2}{2} - \pi(\pi)$$
$$= \pi^2 - \pi^2$$

$$\boxed{a_0 = 0}$$

Solution for Fourier series

$$f(x) = \frac{a_0}{2} + \sum_{n=1}^{\infty} a_n \cos nx + \sum_{n=1}^{\infty} b_n \sin nx$$
$$= \frac{0}{2} - \frac{4}{\pi n^2}\left(\cos x + \cos 3x + \cdots\right)$$
$$+ 0\left(\cos 2x + \cos 4x + \cdots\right)$$
$$+ \frac{2}{n}\left(\sin x + \sin 3x + \cdots\right)$$
$$+ 0\left(\sin 2x + \sin 4x + \cdots\right)$$

4.

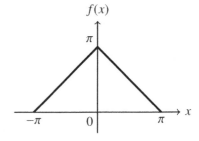

Solution for analytical definition

$$f(x) = \begin{cases} x + \pi & -\pi < x < 0 \\ -x + \pi & 0 < x < \pi \end{cases}$$

Solution for b_n

$$b_n\pi = \text{sum of cases}$$
$$= (x + \pi)(-\pi, 0) + (-x + \pi)(0, \pi)$$
$$= x(-\pi, 0) + \pi(-\pi, 0)$$
$$\quad - x(0, \pi) + \pi(0, \pi)$$
$$= -x(0, -\pi) - \pi(0, -\pi)$$
$$\quad - x(0, \pi) + \pi(0, \pi) \qquad \text{(RP1)}$$

apply Rules 5b and 6b

$$= x(0, \pi) - \pi(0, \pi) - x(0, \pi) + \pi(0, \pi)$$

$$b_n\pi = 0$$

$$\boxed{b_n = 0}$$

Solution for a_n   Continuing from RP1,

$a_n\pi = -x(0, -\pi) - \pi(0, -\pi)$
$\qquad - x(0, \pi) + \pi(0, \pi)$

apply Rules 5a and 6a

$\qquad = -x(0, \pi) + \pi(0, \pi) - x(0, \pi) + \pi(0, \pi)$
$\qquad = -2x(0, \pi) + 2\pi(0, \pi) \qquad\qquad \text{(RP2)}$

$\qquad = -2\left[\dfrac{\pi}{n}\sin n\pi + \dfrac{1}{n^2}(\cos n\pi - 1)\right]$
$\qquad\quad + 2\cdot\dfrac{\pi}{n}\sin n\pi$

$a_n\pi = -2\left[0 + \dfrac{1}{n^2}(\cos n\pi - 1)\right] + 0$

$\quad a_n = -\dfrac{2}{\pi n^2}(\cos n\pi - 1)$

Therefore

$$a_n = \begin{cases} \dfrac{4}{\pi n^2} & n \text{ is odd} \\[2mm] 0 & n \text{ is even} \end{cases}$$

Solution for a_0   Continuing from RP2,

$a_0\pi = -2x(0, \pi) + 2\pi(0, \pi)$

$\qquad = -2\cdot\dfrac{\pi^2}{2} + 2\pi(\pi)$

$\qquad = -\pi^2 + 2\pi^2$

$a_0\pi = \pi^2$

$$a_0 = \pi$$

Solution for Fourier series

$f(x) = \dfrac{a_0}{2} + \displaystyle\sum_{n=1}^{\infty} a_n\cos nx + \sum_{n=1}^{\infty} b_n\sin nx$

$\qquad = \dfrac{\pi}{2} + \dfrac{4}{\pi n^2}\Big(\cos x + \cos 3x + \cdots\Big)$

$\qquad\quad + 0\Big(\cos 2x + \cos 4x + \cdots\Big)$

$\qquad\quad + 0\Big(\sin x + \sin 2x + \cdots\Big)$

$$f(x) = \dfrac{\pi}{2} + \dfrac{4}{\pi}\Big(\cos x + \dfrac{1}{3^2}\cos 3x + \cdots\Big)$$

5.

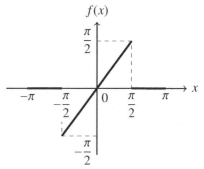

$f(x)$

Solution for analytical definition

$$f(x) = \begin{cases} 0 & -\pi < x < -\dfrac{\pi}{2} \\[2mm] x & -\dfrac{\pi}{2} < x < \dfrac{\pi}{2} \\[2mm] 0 & \dfrac{\pi}{2} < x < \pi \end{cases}$$

Solution for b_n

$b_n\pi = \text{sum of cases}$

$\qquad = 0(-\pi, -\dfrac{\pi}{2}) + x(-\dfrac{\pi}{2}, \dfrac{\pi}{2}) + 0(\dfrac{\pi}{2}, \pi)$

$\qquad = x(-\dfrac{\pi}{2}, \dfrac{\pi}{2})$

$\qquad = x(-\dfrac{\pi}{2}, 0) + x(0, \dfrac{\pi}{2})$

$\qquad = -x(0, -\dfrac{\pi}{2}) + x(0, \dfrac{\pi}{2}) \qquad\qquad \text{(RP1)}$

apply Rule 6b

$\qquad = x(0, \dfrac{\pi}{2}) + x(0, \dfrac{\pi}{2})$

$\qquad = 2x(0, \dfrac{\pi}{2})$

$\qquad = 2\left[-\dfrac{\pi/2}{n}\cos\dfrac{n\pi}{2} + \dfrac{1}{n^2}\sin\dfrac{n\pi}{2}\right]$

$b_n\pi = -\dfrac{\pi}{n}\cos\dfrac{n\pi}{2} + \dfrac{2}{n^2}\sin\dfrac{n\pi}{2}$

$b_n = -\dfrac{1}{n}\cos\dfrac{n\pi}{2} + \dfrac{2}{\pi n^2}\sin\dfrac{n\pi}{2}$

Recall that

$$\cos\dfrac{n\pi}{2} = \begin{cases} 0 & n \text{ is odd} \\[2mm] -1 & n = 2, 6, 10, \ldots \\[2mm] 1 & n = 4, 8, 12, \ldots \end{cases}$$

and

$$\sin \frac{n\pi}{2} = \begin{cases} 0 & n \text{ is even} \\ 1 & n = 1, 5, 9, \ldots \\ -1 & n = 3, 7, 11, \ldots \end{cases}$$

When $n = 1, 5, 9, \ldots$

$$b_n = -0 + \frac{2}{\pi n^2}(1)$$

$$= \frac{2}{\pi n^2}$$

When $n = 2, 6, 10, \ldots$

$$b_n = -\frac{1}{n}(-1) + \frac{2}{\pi n^2}(0)$$

$$= \frac{1}{n}$$

When $n = 3, 7, 11, \ldots$

$$b_n = -\frac{1}{n}(0) + \frac{2}{\pi n^2}(-1)$$

$$= -\frac{2}{\pi n^2}$$

When $n = 4, 8, 12, \ldots$

$$b_n = -\frac{1}{n}(1) + \frac{2}{\pi n^2}(0)$$

$$= -\frac{1}{n}$$

Therefore

$$b_n = \begin{cases} \dfrac{2}{\pi n^2} & n = 1, 5, 9, \ldots \\ \dfrac{1}{n} & n = 2, 6, 10, \ldots \\ -\dfrac{2}{\pi n^2} & n = 3, 7, 11, \ldots \\ -\dfrac{1}{n} & n = 4, 8, 12, \ldots \end{cases}$$

Solution for a_n Continuing from RP1,

$$a_n \pi = -x(0, -\frac{\pi}{2}) + x(0, \frac{\pi}{2})$$

apply Rule 6a

$$= -x(0, \frac{\pi}{2}) + x(0, \frac{\pi}{2})$$

$$= 0$$

$$\boxed{a_n = 0} \qquad \text{(RP2)}$$

Solution for a_0 Continuing from RP2,

$$\boxed{a_0 = 0}$$

Solution for Fourier series

$$f(x) = \frac{a_0}{2} + \sum_{n=1}^{\infty} a_n \cos nx + \sum_{n=1}^{\infty} b_n \sin nx$$

$$= \frac{0}{2} + 0\left(\cos x + \cos 2x + \cdots \right)$$

$$+ b_n\left(\sin x + \sin 5x + \cdots \right)$$

$$+ b_n\left(\sin 2x + \sin 6x + \cdots \right)$$

$$+ b_n\left(\sin 3x + \sin 7x + \cdots \right)$$

$$+ b_n\left(\sin 4x + \sin 8x + \cdots \right)$$

$$= \frac{2}{\pi n^2}\left(\sin x + \sin 5x + \cdots \right)$$

$$+ \frac{1}{n}\left(\sin 2x + \sin 6x + \cdots \right)$$

$$- \frac{2}{\pi n^2}\left(\sin 3x + \sin 7x + \cdots \right)$$

$$- \frac{1}{n}\left(\sin 4x + \sin 8x + \cdots \right)$$

$$= \frac{2}{\pi n^2}\left(\sin x - \sin 3x + \cdots \right)$$

$$+ \frac{1}{n}\left(\sin 2x - \sin 4x + \cdots \right)$$

$$\boxed{\begin{array}{l} f(x) = \dfrac{2}{\pi}\left(\sin x - \dfrac{1}{3^2} \sin 3x + \cdots \right) \\ \qquad + \left(\dfrac{1}{2} \sin 2x - \dfrac{1}{4} \sin 4x + \cdots \right) \end{array}}$$

Chapter 5

1.

$$f(x) = \begin{cases} 0 & -1 < x < 0 \\ x & 0 < x < 1 \end{cases}$$

$$f(x+2) = f(x)$$

Solution for ω   The period T is given as 2.

$$\omega = \frac{2\pi}{T}$$

$$= \frac{2\pi}{2} = \pi$$

Solution for b_n

$$b_n \frac{\pi}{\omega} = \text{sum of cases}$$

put $\omega = \pi$

$$b_n = 0(-1, 0) + x(0, 1)$$
$$= x(0, 1) \tag{RP1}$$

apply Rule 4b

$$= -\frac{1}{n\omega} \cos n\omega(1) + \frac{1}{n^2\omega^2} \sin n\omega(1)$$

$$= -\frac{1}{n\pi} \cos n\pi + \frac{1}{n^2\pi^2} \sin n\pi$$

$$= -\frac{1}{n\pi} \cos n\pi$$

Therefore

$$b_n = \begin{cases} \dfrac{1}{n\pi} & n \text{ is odd} \\ -\dfrac{1}{n\pi} & n \text{ is even} \end{cases}$$

Solution for a_n   Continuing from RP1,

$$a_n = x(0, 1) \tag{RP2}$$

apply Rule 4a

$$= -\frac{1}{n\omega} \sin n\omega(1) + \frac{1}{n^2\omega^2}(\cos n\omega(1) - 1)$$

put $\omega = \pi$

$$= -\frac{1}{n\omega} \sin n\pi + \frac{1}{n^2\pi^2}(\cos n\pi - 1)$$

$$= \frac{1}{n^2\pi^2}(\cos n\pi - 1)$$

Therefore

$$a_n = \begin{cases} -\dfrac{2}{n^2\pi^2} & n \text{ is odd} \\ 0 & n \text{ is even} \end{cases}$$

Solution for a_0   Continuing from RP2,

$$a_0 = x(0, 1)$$

$$= \frac{1^2}{2}$$

$$a_0 = \frac{1}{2}$$

Solution for Fourier series

$$f(x) = \frac{a_0}{2} + \sum_{n=1}^{\infty} a_n \cos n\omega x + \sum_{n=1}^{\infty} b_n \sin n\omega x$$

$$= \frac{1}{4} + a_n(\cos \pi x + \cos 3\pi x + \cdots)$$
$$+ a_n(\cos 2\pi x + \cos 4\pi x + \cdots)$$
$$+ b_n(\sin \pi x + \sin 3\pi x + \cdots)$$
$$+ b_n(\sin 2\pi x + \sin 4\pi x + \cdots)$$

$$= \frac{1}{4} - \frac{2}{n^2\pi^2}(\cos \pi x + \cos 3\pi x + \cdots)$$
$$+ 0(\cos 2\pi x + \cos 4\pi x + \cdots)$$
$$+ \frac{1}{n\pi}(\sin \pi x + \sin 3\pi x + \cdots)$$
$$- \frac{1}{n\pi}(\sin 2\pi x + \sin 4\pi x + \cdots)$$

$$= \frac{1}{4} - \frac{2}{n^2\pi^2}(\cos \pi x + \cos 3\pi x + \cdots)$$
$$+ \frac{1}{n\pi}(\sin \pi x - \sin 2\pi x + \cdots)$$

$$\boxed{\begin{aligned} &f(x) \\ &= \frac{1}{4} - \frac{2}{\pi^2}\left(\cos \pi x + \frac{1}{3^2} \cos 3\pi x + \cdots\right) \\ &+ \frac{1}{\pi}\left(\sin \pi x - \frac{1}{2} \sin 2\pi x + \cdots\right) \end{aligned}}$$

2.

$$f(x) = \begin{cases} 1 & -5 < x < 0 \\ 1 + x & 0 < x < 5 \end{cases}$$

$f(x + 10) = f(x)$

Solution for ω   The period T is given as 10.

$$\omega = \frac{2\pi}{T}$$

$$= \frac{2\pi}{10} = \frac{\pi}{5}$$

Solution for b_n

$b_n \dfrac{\pi}{\omega} = $ sum of cases

put $\omega = \dfrac{\pi}{5}$

$5b_n = 1(-5, 0) + (1 + x)(0, 5)$

$\quad = -(0, -5) + (0, 5)$

$\quad\quad + x(0, 5)$ (RP1)

apply Rule 5b

$\quad = -(0, 5) + (0, 5) + x(0, 5)$

$\quad = x(0, 5)$

$\quad = -\dfrac{5}{n\omega} \cos n\omega(5) + \dfrac{1}{n^2\omega^2} \sin n\omega(5)$

$5b_n = -\dfrac{25}{n\pi} \cos n\pi + \dfrac{25}{n^2\pi^2} \sin n\pi$

$b_n = -\dfrac{5}{n\pi} \cos n\pi$

Therefore

$$b_n = \begin{cases} \dfrac{5}{n\pi} & n \text{ is odd} \\ -\dfrac{5}{n\pi} & n \text{ is even} \end{cases}$$

Solution for a_n   Continuing from RP1,

$5a_n = -(0, -5) + (0, 5) + x(0, 5)$

apply Rule 5a

$\quad = (0, 5) + (0, 5) + x(0, 5)$

$= 2(0, 5) + x(0, 5)$ (RP2)

apply Rules 3a and 4a

$= \dfrac{2}{n\omega} \sin n\omega(5)$

$\quad + \dfrac{5}{n\omega} \sin n\omega(5) + \dfrac{1}{n^2\omega^2}(\cos n\omega(5) - 1)$

$5a_n = \dfrac{10}{n\pi} \sin n\pi$

$\quad + \dfrac{25}{n\pi} \sin n\pi + \dfrac{25}{n^2\pi^2}(\cos n\pi - 1)$

$a_n = \dfrac{5}{n^2\pi^2}(\cos n\pi - 1)$

Therefore

$$a_n = \begin{cases} -\dfrac{10}{n^2\pi^2} & n \text{ is odd} \\ 0 & n \text{ is even} \end{cases}$$

Solution for a_0   Continuing from RP2,

$5a_0 = 2(0, 5) + x(0, 5)$

$\quad = 2(5) + \dfrac{5^2}{2}$

$a_0 = 2 + \dfrac{5}{2}$

$$a_0 = \dfrac{9}{2}$$

Solution for Fourier series

$$f(x) = \frac{a_0}{2} + \sum_{n=1}^{\infty} a_n \cos n\omega x + \sum_{n=1}^{\infty} b_n \sin n\omega x$$

$$= \frac{9}{4} - \frac{10}{n^2\pi^2}\left(\cos\frac{\pi x}{5} + \cos 3\frac{\pi x}{5} + \cdots\right)$$

$$+ 0\left(\cos 2\frac{\pi x}{5} + \cos 4\frac{\pi x}{5} + \cdots\right)$$

$$+ \frac{5}{n\pi}\left(\sin\frac{\pi x}{5} + \sin 3\frac{\pi x}{5} + \cdots\right)$$

$$- \frac{5}{n\pi}\left(\sin 2\frac{\pi x}{5} + \sin 4\frac{\pi x}{5} + \cdots\right)$$

$$= \frac{9}{4} - \frac{10}{n^2\pi^2}\left(\cos\frac{\pi x}{5} + \cos 3\frac{\pi x}{5} + \cdots\right)$$

$$+ \frac{5}{n\pi}\left(\sin\frac{\pi x}{5} - \sin 2\frac{\pi x}{5} + \cdots\right)$$

$$f(x)$$
$$= \frac{9}{4} - \frac{10}{\pi^2}\left(\cos\frac{\pi x}{5} + \frac{1}{3^2}\cos 3\frac{\pi x}{5} + \cdots\right)$$
$$+ \frac{5}{\pi}\left(\sin\frac{\pi x}{5} - \frac{1}{2}\sin 2\frac{\pi x}{5} + \cdots\right)$$

3.

$$f(x) = \begin{cases} 2 + x & -2 < x < 0 \\ 2 & 0 < x < 2 \end{cases}$$

$$f(x+4) = f(x)$$

Solution for ω   The period T is given as 4.

$$\omega = \frac{2\pi}{T}$$
$$= \frac{2\pi}{4} = \frac{\pi}{2}$$

Solution for b_n

$$b_n\frac{\pi}{\omega} = \text{sum of cases}$$

put $\omega = \dfrac{\pi}{2}$

$$2b_n = (2+x)(-2,0) + 2(0,2)$$
$$= 2(-2,0) + x(-2,0) + 2(0,2)$$
$$= -2(0,-2) - x(0,-2)$$
$$\qquad + 2(0,2) \qquad\qquad \text{(RP1)}$$

apply Rules 5b and 6b

$$= -2(0,2) + x(0,2) + 2(0,2)$$
$$= x(0,2)$$
$$= -\frac{2}{n\omega}\cos n\omega(2) + \frac{1}{n^2\omega^2}\sin n\omega(2)$$
$$2b_n = -\frac{4}{n\pi}\cos n\pi + \frac{4}{n^2\pi^2}\sin n\pi$$
$$b_n = -\frac{2}{n\pi}\cos n\pi$$

Therefore

$$b_n = \begin{cases} \dfrac{2}{n\pi} & n \text{ is odd} \\[2mm] -\dfrac{2}{n\pi} & n \text{ is even} \end{cases}$$

Solution for a_n   Continuing from RP1,

$$2a_n = -2(0,-2) - x(0,-2) + 2(0,2)$$

apply Rules 5b and 6b

$$= 2(0,2) - x(0,2) + 2(0,2)$$
$$= 4(0,2) - x(0,2) \qquad\qquad \text{(RP2)}$$

apply Rules 3a and 4a

$$= \frac{8}{n\omega}\sin n\omega(2) - \left[\frac{2}{n\omega}\sin n\omega(2)\right.$$
$$\left. + \frac{1}{n^2\omega^2}(\cos n\omega(2) - 1)\right]$$
$$2a_n = \frac{4}{n\pi}\sin n\pi - \frac{4}{n\pi}\sin n\pi$$
$$\qquad - \frac{4}{n^2\pi^2}(\cos n\pi - 1)$$
$$a_n = -\frac{2}{n^2\pi^2}(\cos n\pi - 1)$$

Therefore

$$a_n = \begin{cases} \dfrac{4}{n^2\pi^2} & n \text{ is odd} \\[2mm] 0 & n \text{ is even} \end{cases}$$

Solution for a_0   Continuing from RP2,

$$2a_0 = 4(0,2) - x(0,2)$$
$$= 4(2) - \frac{2^2}{2}$$
$$a_0 = 4 - \frac{2}{2}$$
$$\boxed{a_0 = 3}$$

Solution for Fourier series

$$f(x) = \frac{a_0}{2} + \sum_{n=1}^{\infty} a_n \cos n\omega x + \sum_{n=1}^{\infty} b_n \sin n\omega x$$

$$= \frac{3}{2} + \frac{4}{n^2\pi^2}\left(\cos\frac{\pi x}{2} + \cos 3\frac{\pi x}{2} + \cdots\right)$$

$$+ 0\left(\cos 2\frac{\pi x}{2} + \cos 4\frac{\pi x}{2} + \cdots\right)$$

$$+ \frac{2}{n\pi}\left(\sin\frac{\pi x}{2} + \sin 3\frac{\pi x}{2} + \cdots\right)$$

$$- \frac{2}{n\pi}\left(\sin 2\frac{\pi x}{2} + \sin 4\frac{\pi x}{2} + \cdots\right)$$

$$= \frac{3}{2} + \frac{4}{n^2\pi^2}\left(\cos\frac{\pi x}{2} + \cos 3\frac{\pi x}{2} + \cdots\right)$$

$$+ \frac{2}{n\pi}\left(\sin\frac{\pi x}{2} - \sin 2\frac{\pi x}{2} + \cdots\right)$$

$$\boxed{\begin{aligned} f(x) &= \frac{3}{2} + \frac{4}{\pi^2}\left(\cos\frac{\pi x}{2} + \frac{1}{3^2}\cos 3\frac{\pi x}{2} + \cdots\right) \\ &+ \frac{2}{\pi}\left(\sin\frac{\pi x}{2} - \frac{1}{2}\sin 2\frac{\pi x}{2} + \cdots\right) \end{aligned}}$$

4.

$$f(x) = \begin{cases} x + 5 & -2 < x < 0 \\ -x + 5 & 0 < x < 2 \end{cases}$$

$$f(x + 4) = f(x)$$

Solution for ω   The period T is given as 4.

$$\omega = \frac{2\pi}{T}$$

$$= \frac{2\pi}{4} = \frac{\pi}{2}$$

Solution for b_n

$$b_n\frac{\pi}{\omega} = \text{sum of cases}$$

put $\omega = \frac{\pi}{2}$

$$2b_n = (x+5)(-2, 0) + (-x+5)(0, 2)$$

$$= x(-2, 0) + 5(-2, 0)$$

$$- x(0, 2) + 5(0, 2)$$

$$= -x(0, -2) - 5(0, -2)$$
$$- x(0, 2) + 5(0, 2) \tag{RP1}$$

apply Rules 5b and 6b

$$= x(0, 2) - 5(0, 2) - x(0, 2) + 5(0, 2)$$

$$2b_n = 0$$

$$\boxed{b_n = 0}$$

Solution for a_n   Continuing from RP1,

$$2a_n = -x(0, -2) - 5(0, -2)$$
$$- x(0, 2) + 5(0, 2)$$

apply Rules 5b and 6b

$$= -x(0, 2) + 5(0, 2) - x(0, 2) + 5(0, 2)$$
$$= -2x(0, 2) + 10(0, 2) \tag{RP2}$$

apply Rules 3a and 4a

$$= -2\left[\frac{2}{n\omega}\sin n\omega(2)\right.$$
$$\left. + \frac{1}{n^2\omega^2}(\cos n\omega(2) - 1)\right]$$
$$+ 10\frac{1}{n\omega}\sin n\omega(2)$$

$$= -2\left[\frac{4}{n\pi}\sin n\pi + \frac{4}{n^2\pi^2}(\cos n\pi - 1)\right]$$
$$+ \frac{20}{n\pi}\sin n\pi$$

$$2a_n = -2\left[0 + \frac{4}{n^2\pi^2}(\cos n\pi - 1)\right]$$

$$a_n = -\frac{4}{n^2\pi^2}(\cos n\pi - 1)$$

Therefore

$$\boxed{a_n = \begin{cases} \dfrac{8}{n^2\pi^2} & n \text{ is odd} \\ 0 & n \text{ is even} \end{cases}}$$

Solution for a_0   Continuing from RP2,

$$2a_n = -2x(0, 2) + 10(0, 2)$$

$$= -2\left(\frac{2^2}{2}\right) + 10(2)$$

$$2a_0 = 16$$

$$\boxed{a_0 = 8}$$

Solution for Fourier series

$$f(x) = \frac{a_0}{2} + \sum_{n=1}^{\infty} a_n \cos n\omega x + \sum_{n=1}^{\infty} b_n \sin n\omega x$$

$$= \frac{8}{2} + \frac{8}{n^2\pi^2}\left(\cos \frac{\pi x}{2} + \cos 3\frac{\pi x}{2} + \cdots \right)$$

$$+ 0\left(\cos 2\frac{\pi x}{2} + \cos 4\frac{\pi x}{2} + \cdots \right)$$

$$+ 0\left(\sin \frac{\pi x}{2} + \sin 2\frac{\pi x}{2} + \cdots \right)$$

$$\boxed{f(x) = 4 + \frac{8}{\pi^2}\left(\cos \frac{\pi x}{2} + \frac{1}{3^2}\cos 3\frac{\pi x}{2} + \cdots \right)}$$

5.

$$f(x) = \begin{cases} \pi & -1 < x < 0 \\ -\pi & 0 < x < 1 \end{cases}$$

$$f(x + 2) = f(x)$$

Solution for ω   The period T is given as 2.

$$\omega = \frac{2\pi}{T}$$

$$= \frac{2\pi}{2} = \pi$$

Solution for b_n

$$b_n \frac{\pi}{\omega} = \text{sum of cases}$$

put $\omega = \pi$

$$b_n = \pi(-1, 0) - \pi(0, 1)$$

$$= -\pi(0, -1) - \pi(0, 1) \qquad \text{(RP1)}$$

apply Rule 5b

$$= -\pi(0, 1) - \pi(0, 1)$$

$$= -2\pi(0, 1)$$

$$= -2\pi\frac{1}{n\omega}[1 - \cos n\omega(1)]$$

$$= -2\pi\frac{1}{n\pi}(1 - \cos n\pi)$$

$$= -\frac{2}{n}(1 - \cos n\pi)$$

Therefore

$$\boxed{b_n = \begin{cases} -\dfrac{4}{n} & n \text{ is odd} \\ 0 & n \text{ is even} \end{cases}}$$

Solution for a_n   Continuing from RP1,

$$a_n = -\pi(0, -1) - \pi(0, 1)$$

apply Rule 5a

$$= \pi(0, 1) - \pi(0, 1)$$

$$\boxed{a_n = 0} \qquad \text{(RP2)}$$

Solution for a_0   Continuing from RP2,

$$\boxed{a_0 = 0}$$

Solution for Fourier series

$$f(x) = \frac{a_0}{2} + \sum_{n=1}^{\infty} a_n \cos n\omega x + \sum_{n=1}^{\infty} b_n \sin n\omega x$$

$$= \frac{0}{2} + 0(\cos \pi x + \cos 3\pi x + \cdots)$$

$$+ b_n(\sin \pi x + \sin 3\pi x + \cdots)$$

$$+ b_n(\sin 2\pi x + \sin 4\pi x + \cdots)$$

$$= -\frac{4}{n}(\sin \pi x + \sin 3\pi x + \cdots)$$

$$+ 0(\sin 2\pi x + \sin 4\pi x + \cdots)$$

$$\boxed{f(x) = -4\left(\sin \pi x + \frac{1}{3}\sin 3\pi x + \cdots \right)}$$

6.

$$f(x) = 1 - x \quad 0 < x < 2$$

$$f(x + 2) = f(x)$$

Solution for ω The period T is given as 2.

$$\omega = \frac{2\pi}{T}$$
$$= \frac{2\pi}{2} = \pi$$

Solution for b_n

$$b_n\frac{\pi}{\omega} = \text{sum of cases}$$

put $\omega = \pi$

$$b_n = (1 - x)(0, 2)$$
$$= (0, 2) - x(0, 2) \qquad \text{(RP1)}$$
$$= \frac{1}{n\omega}[1 - \cos n\omega(2)]$$
$$\quad - \left[-\frac{2}{n\omega} \cos n\omega(2) \right.$$
$$\quad + \left. \frac{1}{n^2\omega^2} \sin n\omega(2) \right]$$
$$= \frac{1}{n\pi}(1 - \cos 2n\pi)$$
$$\quad + \frac{2}{n\pi} \cos 2n\pi - \frac{1}{n^2\pi^2} \sin 2n\pi$$
$$= \frac{1}{n\pi}(1 - 1) + \frac{2}{n\pi}(1) - 0$$

$$\boxed{b_n = \frac{2}{n\pi}}$$

Solution for a_n Continuing from RP1,

$$a_n = (0, 2) - x(0, 2) \qquad \text{(RP2)}$$
$$= \frac{1}{n\omega} \sin n\omega(2) - \left[\frac{2}{n\omega} \sin n\omega(2) \right.$$
$$\quad + \left. \frac{1}{n^2\omega^2}(\cos n\omega(2) - 1) \right]$$
$$= \frac{1}{n\pi} \sin 2n\pi - \left[\frac{2}{n\pi} \sin 2n\pi \right.$$
$$\quad + \left. \frac{1}{n^2\pi^2}(\cos 2n\pi - 1) \right]$$
$$= 0 - \left[0 + \frac{1}{n^2\pi^2}(1 - 1) \right]$$

$$\boxed{a_n = 0}$$

Solution for a_0 Continuing from RP2,

$$a_0 = (0, 2) - x(0, 2)$$
$$= 2 - \frac{2^2}{2}$$

$$\boxed{a_0 = 0}$$

Solution for Fourier series

$$f(x) = \frac{a_0}{2} + \sum_{n=1}^{\infty} a_n \cos n\omega x + \sum_{n=1}^{\infty} b_n \sin n\omega x$$
$$= \frac{0}{2} + 0(\cos \pi x + \cos 3\pi x + \cdots)$$
$$\quad + b_n(\sin \pi x + \sin 2\pi x + \cdots)$$
$$= \frac{2}{n\pi}(\sin \pi x + \sin 2\pi x + \cdots)$$

$$\boxed{f(x) = \frac{2}{\pi}\left(\sin \pi x + \frac{1}{2} \sin 2\pi x + \cdots\right)}$$

7.

$$f(x) = \begin{cases} 1 & -2 < x < -1 \\ 0 & -1 < x < 1 \\ 1 & 1 < x < 2 \end{cases}$$
$$f(x + 4) = f(x)$$

Solution for ω The period T is given as 4.

$$\omega = \frac{2\pi}{T}$$
$$= \frac{2\pi}{4} = \frac{\pi}{2}$$

Solution for b_n

$$b_n\frac{\pi}{\omega} = \text{sum of cases}$$

put $\omega = \frac{\pi}{2}$

$$2b_n = 1(-2, -1) + 0(-1, 1) + 1(1, 2)$$
$$= (-2, -1) + (1, 2)$$
$$= -(-1, -2) + (1, 2) \qquad \text{(RP1)}$$

apply Rule 5b

$$2b_n = -(1,2) + (1,2)$$

$$\boxed{b_n = 0}$$

Solution for a_n   Continuing from RP1,

$$2a_n = -(-1,-2) + (1,2)$$

apply Rule 5a

$$2a_n = (1,2) + (1,2)$$

$$a_n = (1,2)$$

$$= (1,0) + (0,2)$$

$$= -(0,1) + (0,2) \qquad \text{(RP2)}$$

$$= -\frac{1}{n\omega}\sin n\omega(1) + \frac{1}{n\omega}\sin n\omega(2)$$

$$= -\frac{2}{n\pi}\sin\frac{n\pi}{2} + \frac{2}{n\pi}\sin n\pi$$

$$a_n = -\frac{2}{n\pi}\sin\frac{n\pi}{2}$$

Recall that

$$\sin\frac{n\pi}{2} = \begin{cases} 0 & n \text{ is even} \\ 1 & n = 1, 5, 9, \ldots \\ -1 & n = 3, 7, 11, \ldots \end{cases}$$

Therefore

$$a_n = \begin{cases} 0 & n \text{ is even} \\ -\dfrac{2}{n\pi} & n = 1, 5, 9, \ldots \\ \dfrac{2}{n\pi} & n = 3, 7, 11, \ldots \end{cases}$$

Solution for a_0   Continuing from RP2,

$$a_0 = -(0,1) + (0,2)$$

$$= -1 + 2$$

$$\boxed{a_0 = 1}$$

Solution for Fourier series

$$f(x) = \frac{a_0}{2} + \sum_{n=1}^{\infty} a_n \cos n\omega x + \sum_{n=1}^{\infty} b_n \sin n\omega x$$

$$= \frac{1}{2} + a_n\left(\cos 2\frac{\pi x}{2} + \cos 4\frac{\pi x}{2} + \cdots\right)$$

$$+ a_n\left(\cos\frac{\pi x}{2} + \cos 5\frac{\pi x}{2} + \cdots\right)$$

$$+ a_n\left(\cos 3\frac{\pi x}{2} + \cos 7\frac{\pi x}{2} + \cdots\right)$$

$$+ 0\left(\sin\frac{\pi x}{2} + \sin 2\frac{\pi x}{2} + \cdots\right)$$

$$= \frac{1}{2} + 0\left(\cos 2\frac{\pi x}{2} + \cos 4\frac{\pi x}{2} + \cdots\right)$$

$$- \frac{2}{n\pi}\left(\cos\frac{\pi x}{2} + \cos 5\frac{\pi x}{2} + \cdots\right)$$

$$+ \frac{2}{n\pi}\left(\cos 3\frac{\pi x}{2} + \cos 7\frac{\pi x}{2} + \cdots\right)$$

$$= \frac{1}{2} - \frac{2}{n\pi}\left(\cos\frac{\pi x}{2} - \cos 3\frac{\pi x}{2} + \cdots\right)$$

$$\boxed{f(x) = \frac{1}{2} - \frac{2}{\pi}\left(\cos\frac{\pi x}{2} - \frac{1}{3}\cos 3\frac{\pi x}{2} + \cdots\right)}$$

8.

$$f(x) = \begin{cases} x + 1 & -1 < x < 0 \\ x - 1 & 0 < x < 1 \end{cases}$$

$$f(x + 2) = f(x)$$

Solution for ω   The period T is given as 2.

$$\omega = \frac{2\pi}{T}$$

$$= \frac{2\pi}{2} = \pi$$

Solution for b_n

$$b_n\frac{\pi}{\omega} = \text{sum of cases}$$

put $\omega = \pi$

$$b_n = (x + 1)(-1,0) + (x - 1)(0,1)$$

$$= x(-1,0) + (-1,0) + x(0,1) - (0,1)$$

$$= -x(0,-1) - (0,-1)$$

$$+ x(0,1) - (0,1) \qquad \text{(RP1)}$$

apply Rules 5b and 6b

$$= x(0,1) - (0,1) + x(0,1) - (0,1)$$

$$= 2x(0,1) - 2(0,1)$$

$$= 2\left[-\frac{1}{n\omega}\cos n\omega(1) + \frac{1}{n^2\omega^2}\sin n\omega(1)\right]$$

$$-\frac{2}{n\omega}[1 - \cos n\omega(1)]$$

$$= 2\left[-\frac{1}{n\pi}\cos n\pi + \frac{1}{n^2\pi^2}\sin n\pi\right]$$

$$-\frac{2}{n\pi}(1 - \cos n\pi)$$

$$= 2\left[-\frac{1}{n\pi}\cos n\pi + 0\right] - \frac{2}{n\pi}(1 - \cos n\pi)$$

$$b_n = -\frac{2}{n\pi}\cos n\pi - \frac{2}{n\pi}(1) + \frac{2}{n\pi}(\cos n\pi)$$

$$\boxed{b_n = -\frac{2}{n\pi}}$$

Solution for a_n   Continuing from RP1,

$$a_n = -x(0, -1) - (0, -1) + x(0, 1) - (0, 1)$$

apply Rules 5a and 6a

$$= -x(0, 1) + (0, 1) + x(0, 1) - (0, 1)$$

$$\boxed{a_n = 0} \qquad\qquad \text{(RP2)}$$

Solution for a_0   Continuing from RP2,

$$\boxed{a_0 = 0}$$

Solution for Fourier series

$$f(x) = \frac{a_0}{2} + \sum_{n=1}^{\infty} a_n \cos n\omega x + \sum_{n=1}^{\infty} b_n \sin n\omega x$$

$$= \frac{0}{2} + 0(\cos \pi x + \cos 3\pi x + \cdots)$$

$$+ b_n(\sin \pi x + \sin 2\pi x + \cdots)$$

$$= -\frac{2}{n\pi}(\sin \pi x + \sin 2\pi x + \cdots)$$

$$\boxed{f(x) = -\frac{2}{\pi}\left(\sin \pi x + \frac{1}{2}\sin 2\pi x + \cdots\right)}$$

9.

$$f(x) = \begin{cases} 1 & -2 < x < -1 \\ -x & -1 < x < 0 \\ x & 0 < x < 1 \\ 1 & 1 < x < 2 \end{cases}$$

$$f(x + 4) = f(x)$$

Solution for ω   The period T is given as 4.

$$\omega = \frac{2\pi}{T}$$

$$= \frac{2\pi}{4} = \frac{\pi}{2}$$

Solution for b_n

$$b_n\frac{\pi}{\omega} = \text{sum of cases}$$

put $\omega = \dfrac{\pi}{2}$

$$2b_n = (-2, -1) - x(-1, 0) + x(0, 1) + (1, 2)$$

$$= -(-1, -2) + x(0, -1)$$

$$+ x(0, 1) + (1, 2) \qquad\qquad \text{(RP1)}$$

apply Rules 5b and 6b

$$= -(1, 2) - x(0, 1) + x(0, 1) + (1, 2)$$

$$2b_n = 0$$

$$\boxed{b_n = 0}$$

Solution for a_n   Continuing from RP1,

$$2a_n = -(-1, -2) + x(0, -1)$$

$$+ x(0, 1) + (1, 2)$$

apply Rules 5a and 6a

$$= (1, 2) + x(0, 1) + x(0, 1) + (1, 2)$$

$$= 2(1, 2) + 2x(0, 1)$$

$$= 2(1, 0) + 2(0, 2) + 2x(0, 1)$$

$$= -2(0, 1) + 2(0, 2)$$

$$+ 2x(0, 1) \qquad\qquad \text{(RP2)}$$

$$= -2\frac{1}{n\omega}\sin n\omega(1) + 2\frac{1}{n\omega}\sin n\omega(2)$$

$$+ 2\left[\frac{1}{n\omega}\sin n\omega(1)\right.$$

$$\left. + \frac{1}{n^2\omega^2}(\cos n\omega(1) - 1)\right]$$

$$= -\frac{4}{n\pi}\sin\frac{n\pi}{2} + \frac{4}{n\pi}\sin n\pi$$

$$+ \frac{4}{n\pi}\sin\frac{n\pi}{2} + \frac{8}{n^2\pi^2}(\cos\frac{n\pi}{2} - 1)$$

$$2a_n = \frac{8}{n^2\pi^2}(\cos\frac{n\pi}{2} - 1)$$

$$a_n = \frac{4}{n^2\pi^2}(\cos\frac{n\pi}{2} - 1)$$

Recall that

$$\cos\frac{n\pi}{2} = \begin{cases} 0 & n \text{ is odd} \\ -1 & n = 2, 6, 10, \ldots \\ 1 & n = 4, 8, 12, \ldots \end{cases}$$

Therefore

$$a_n = \begin{cases} -\dfrac{4}{n^2\pi^2} & n \text{ is odd} \\ -\dfrac{8}{n^2\pi^2} & n = 2, 6, 10, \ldots \\ 0 & n = 4, 8, 12, \ldots \end{cases}$$

Solution for a_0 Continuing from RP2,

$$2a_n = -2(0, 1) + 2(0, 2) + 2x(0, 1)$$

$$= -2(1) + 2(2) + 2\left(\frac{1^2}{2}\right)$$

$$= -2 + 4 + 1$$

$$2a_0 = 3$$

$$\boxed{a_0 = \frac{3}{2}}$$

Solution for Fourier series

$$f(x) = \frac{a_0}{2} + \sum_{n=1}^{\infty} a_n \cos n\omega x + \sum_{n=1}^{\infty} b_n \sin n\omega x$$

$$= \frac{3}{4} + a_n(\cos\frac{\pi x}{2} + \cos 3\frac{\pi x}{2} + \cdots)$$

$$+ a_n(\cos 2\frac{\pi x}{2} + \cos 6\frac{\pi x}{2} + \cdots)$$

$$+ a_n(\cos 4\frac{\pi x}{2} + \cos 8\frac{\pi x}{2} + \cdots)$$

$$+ b_n(\sin\frac{\pi x}{2} + \sin 2\frac{\pi x}{2} + \cdots)$$

$$= \frac{3}{4} - \frac{4}{n^2\pi^2}(\cos\frac{\pi x}{2} + \cos 3\frac{\pi x}{2} + \cdots)$$

$$- \frac{8}{n^2\pi^2}(\cos 2\frac{\pi x}{2} + \cos 6\frac{\pi x}{2} + \cdots)$$

$$+ 0(\cos 4\frac{\pi x}{2} + \cos 8\frac{\pi x}{2} + \cdots)$$

$$+ 0(\sin\frac{\pi x}{2} + \sin 2\frac{\pi x}{2} + \cdots)$$

$$= \frac{3}{4} - \frac{4}{n^2\pi^2}(\cos\frac{\pi x}{2} + \cos 3\frac{\pi x}{2} + \cdots)$$

$$- \frac{8}{n^2\pi^2}(\cos 2\frac{\pi x}{2} + \cos 6\frac{\pi x}{2} + \cdots)$$

$$\boxed{\begin{aligned} f(x) &= \frac{3}{4} - \frac{4}{\pi^2}(\cos\frac{\pi x}{2} + \frac{1}{3^2}\cos 3\frac{\pi x}{2} + \cdots) \\ &\quad - \frac{8}{\pi^2}(\frac{1}{2^2}\cos 2\frac{\pi x}{2} + \frac{1}{6^2}\cos 6\frac{\pi x}{2} + \cdots) \end{aligned}}$$

10.

$$f(x) = \begin{cases} 1 & 0 < x < \dfrac{1}{2} \\ 0 & \dfrac{1}{2} < x < 1 \end{cases}$$

$$f(x + 1) = f(x)$$

Solution for ω The period T is given as 1.

$$\omega = \frac{2\pi}{T}$$

$$= \frac{2\pi}{1} = 2\pi$$

Solution for b_n

$$b_n\frac{\pi}{\omega} = \text{sum of cases}$$

put $\omega = 2\pi$

$$\frac{1}{2}b_n = 1(0, \frac{1}{2}) + 0(\frac{1}{2}, 1)$$

$$= (0, \frac{1}{2}) \qquad\qquad \text{(RP1)}$$

$$= \frac{1}{n\omega}[1 - \cos n\omega(\frac{1}{2})]$$

$$\frac{1}{2}b_n = \frac{1}{2n\pi}[1 - \cos 2n\pi(\frac{1}{2})]$$

$$b_n = \frac{1}{n\pi}(1 - \cos n\pi)$$

Therefore

$$b_n = \begin{cases} \dfrac{2}{n\pi} & n \text{ is odd} \\ 0 & n \text{ is even} \end{cases}$$

Solution for a_n   Continuing from RP1,

$$\frac{1}{2}a_n = (0, \frac{1}{2})$$ (RP2)

$$= \frac{1}{n\omega}\sin n\omega(\frac{1}{2})$$

$$= \frac{1}{2n\pi}\sin n\pi$$

$$= 0$$

$$\boxed{a_n = 0}$$

Solution for a_0   Continuing from RP2,

$$\frac{1}{2}a_0 = (0, \frac{1}{2})$$

$$a_0 = 2(\frac{1}{2})$$

$$\boxed{a_0 = 1}$$

Solution for Fourier series

$$f(x) = \frac{a_0}{2} + \sum_{n=1}^{\infty} a_n \cos n\omega x + \sum_{n=1}^{\infty} b_n \sin n\omega x$$

$$= \frac{1}{2} + 0(\cos(2\pi x) + \cos 3(2\pi x) + \cdots)$$

$$+ b_n(\sin(2\pi x) + \sin 3(2\pi x) + \cdots)$$

$$+ b_n(\sin 2(2\pi x) + \sin 4(2\pi x) + \cdots)$$

$$= \frac{1}{2} + \frac{2}{n\pi}(\sin(2\pi x) + \sin 3(2\pi x) + \cdots)$$

$$+ 0(\sin 2(2\pi x) + \sin 4(2\pi x) + \cdots)$$

$$\boxed{f(x) = \frac{1}{2} + \frac{2}{\pi}(\sin(2\pi x) + \frac{1}{3}\sin 3(2\pi x) + \cdots)}$$

Solution for ω   The period T is given as 1.

$$\omega = \frac{2\pi}{T}$$

$$= \frac{2\pi}{1} = 2\pi$$

Solution for b_n

$$b_n \frac{\pi}{\omega} = \text{sum of cases}$$

$$\text{put } \omega = 2\pi$$

$$\frac{1}{2}b_n = 0(0, \frac{1}{2}) + 1(\frac{1}{2}, 1)$$

$$= (\frac{1}{2}, 1)$$

$$= (\frac{1}{2}, 0) + (0, 1)$$

$$= -(0, \frac{1}{2}) + (0, 1)$$ (RP1)

$$= -\frac{1}{n\omega}[1 - \cos n\omega(\frac{1}{2})]$$

$$+ \frac{1}{n\omega}[1 - \cos n\omega(1)]$$

$$\frac{1}{2}b_n = -\frac{1}{2n\pi}(1 - \cos n\pi)$$

$$+ \frac{1}{2n\pi}(1 - \cos 2n\pi)$$

$$b_n = -\frac{1}{n\pi}(1 - \cos n\pi)$$

Therefore

$$b_n = \begin{cases} -\dfrac{2}{n\pi} & n \text{ is odd} \\ 0 & n \text{ is even} \end{cases}$$

11.

$$f(x) = \begin{cases} 0 & 0 < x < \dfrac{1}{2} \\ 1 & \dfrac{1}{2} < x < 1 \end{cases}$$

$$f(x + 1) = f(x)$$

Solution for a_n Continuing from RP1,

$$\frac{1}{2}a_n = -(0, \frac{1}{2}) + (0, 1) \qquad \text{(RP2)}$$

$$= -\frac{1}{n\omega} \sin n\omega(\frac{1}{2}) + \frac{1}{n\omega} \sin n\omega(1)$$

$$= -\frac{1}{2n\pi} \sin n\pi + \frac{1}{2n\pi} \sin 2n\pi$$

$$\frac{1}{2}a_n = 0 + 0$$

$$\boxed{a_n = 0}$$

Solution for a_0 Continuing from RP2,

$$\frac{1}{2}a_0 = -(0, \frac{1}{2}) + (0, 1)$$

$$= -\frac{1}{2} + 1$$

$$= \frac{1}{2}$$

$$\boxed{a_0 = 1}$$

Solution for Fourier series

$$f(x)$$

$$= \frac{a_0}{2} + \sum_{n=1}^{\infty} a_n \cos n\omega x + \sum_{n=1}^{\infty} b_n \sin n\omega x$$

$$= \frac{1}{2} + 0(\cos(2\pi x) + \cos 3(2\pi x) + \cdots)$$

$$+ b_n(\sin(2\pi x) + \sin 3(2\pi x) + \cdots)$$

$$+ b_n(\sin 2(2\pi x) + \sin 4(2\pi x) + \cdots)$$

$$= \frac{1}{2} - \frac{2}{n\pi}(\sin(2\pi x) + \sin 3(2\pi x) + \cdots)$$

$$+ 0(\sin 2(2\pi x) + \sin 4(2\pi x) + \cdots)$$

$$\boxed{\begin{array}{l} f(x) \\ = \dfrac{1}{2} - \dfrac{2}{\pi}(\sin(2\pi x) + \dfrac{1}{3} \sin 3(2\pi x) + \cdots) \end{array}}$$

Chapter 6

Exercise 1

1.

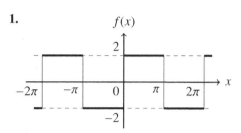

$f(x)$ is *odd*

2.

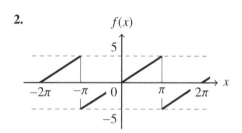

$f(x)$ is *odd*

3.

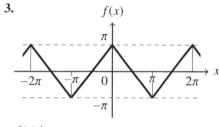

$f(x)$ is *even*

4.

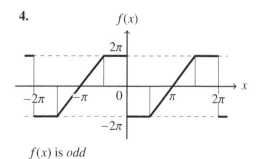

$f(x)$ is *odd*

Exercise 2

1. $f(x) = \dfrac{\pi}{2} + 2\left(\sin x + \dfrac{1}{3} \sin 3x + \cdots \right)$

Solution

$f(x) = \dfrac{\pi}{2} + 2\left(\sin x + \dfrac{1}{3} \sin 3x + \cdots \right)$

$\quad = E + E(D + D + \cdots)$

$\quad = E + E \times D$

$\quad = E + D \qquad \text{where } E \neq 0 \text{ and } D \neq 0$

Therefore $f(x)$ is neither odd nor even.

2. $f(x) = 2\left(\sin x + \dfrac{1}{2} \sin 2x + \cdots \right)$

Solution

$f(x) = 2\left(\sin x + \dfrac{1}{2} \sin 2x + \cdots \right)$

$\quad = E(D + D + \cdots)$

$\quad = E \times D$

$\quad = D$

Therefore $f(x)$ is odd.

3. $f(x) = \dfrac{\pi}{2} - \dfrac{4}{\pi}\left(\cos x + \dfrac{1}{3^2} \cos 3x + \cdots \right)$

Solution

$f(x) = \dfrac{\pi}{2} - \dfrac{4}{\pi}\left(\cos x + \dfrac{1}{3^2} \cos 3x + \cdots \right)$

$\quad = E + E(E + E + \cdots)$

$\quad = E + E \times E$

$\quad = E$

Therefore $f(x)$ is even.

4.

$f(x) = \dfrac{3\pi}{2} - \dfrac{2}{\pi}\left(\cos x + \dfrac{1}{3^2} \cos 3x + \cdots \right)$

$\qquad - \left(\sin x - \dfrac{1}{2} \sin 2x + \cdots \right)$

Solution

$f(x) = \dfrac{3\pi}{2} - \dfrac{2}{\pi}\left(\cos x + \dfrac{1}{3^2} \cos 3x + \cdots \right)$

$\qquad - \left(\sin x - \dfrac{1}{2} \sin 2x + \cdots \right)$

$\quad = E + E(E + E + \cdots) + E(D + D + \cdots)$

$\quad = E + E \times E + E \times D$

$\quad = E + E + D$

$\quad = E + D \qquad \text{where } E \neq 0 \text{ and } D \neq 0$

Therefore $f(x)$ is neither odd nor even.

5.

$f(x) = 2\left(\cos x - \dfrac{1}{3} \cos 3x + \cdots \right)$

$\qquad + \dfrac{4}{\pi}\left(\dfrac{1}{2} \sin 2x + \dfrac{1}{6} \sin 6x + \cdots \right)$

Solution

$= 2\left(\cos x - \dfrac{1}{3} \cos 3x + \cdots \right)$

$\quad + \dfrac{4}{\pi}\left(\dfrac{1}{2} \sin 2x + \dfrac{1}{6} \sin 6x + \cdots \right)$

$= E(E + E + \cdots) + E(D + D + \cdots)$

$= E \times E + E \times D$

$= E + D \qquad \text{where } E \neq 0 \text{ and } D \neq 0$

Therefore $f(x)$ is neither odd nor even.

Exercise 3

1.

$$f(x) = \begin{cases} -4x & -\pi < x < 0 \\ 4x & 0 < x < \pi \end{cases}$$

$$f(x + 2\pi) = f(x)$$

Determine the property of $f(x)$ The graph of $f(x)$ below shows that it is an even function. Thus

$$\boxed{b_n = 0}$$

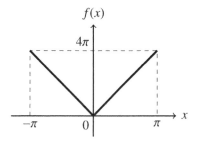

Solution for a_n

$a_n\pi$ = sum of cases

$$= -4x(-\pi, 0) + 4x(0, \pi)$$

$$= 4x(0, -\pi) + 4x(0, \pi)$$

apply Rule 6a

$$= 4x(0, \pi) + 4x(0, \pi)$$

$$= 8x(0, \pi) \qquad\qquad \text{(RP)}$$

apply Rule 4a

$$= 8\left[\frac{1}{n}\sin n\pi + \frac{1}{n^2}(\cos n\pi - 1)\right]$$

$$a_n\pi = 8\left[0 + \frac{1}{n^2}(\cos n\pi - 1)\right]$$

$$a_n = \frac{8}{\pi n^2}(\cos n\pi - 1)$$

Therefore

$$a_n = \begin{cases} -\dfrac{16}{\pi n^2} & n \text{ is odd} \\ 0 & n \text{ is even} \end{cases}$$

Solution for a_0 Continuing from the reuse point,

$$a_0\pi = 8x(0, \pi)$$

$$= 8 \times \frac{\pi^2}{2}$$

$$a_0\pi = 4\pi^2$$

$$\boxed{a_0 = 4\pi}$$

Solution for Fourier series

$$f(x) = \frac{a_0}{2} + \sum_{n=1}^{\infty} a_n \cos nx$$

$$= \frac{4\pi}{2} + a_n(\cos x + \cos 3x + \cdots)$$

$$+ a_n(\cos 2x + \cos 4x + \cdots)$$

$$= 2\pi - \frac{16}{\pi n^2}(\cos x + \cos 3x + \cdots)$$

$$+ 0(\cos 2x + \cos 4x + \cdots)$$

$$\boxed{f(x) = 2\pi - \frac{16}{\pi}\left(\cos x + \frac{1}{3^2}\cos 3x + \cdots\right)}$$

2.

$$f(x) = \begin{cases} -\dfrac{\pi}{4} & -\pi < x < 0 \\ \dfrac{\pi}{4} & 0 < x < \pi \end{cases}$$

$$f(x + 2\pi) = f(x)$$

Determine the property of $f(x)$ The graph of $f(x)$ below shows that it is an odd function. Thus

$$\boxed{a_n = 0} \qquad \text{and} \qquad \boxed{a_0 = 0}$$

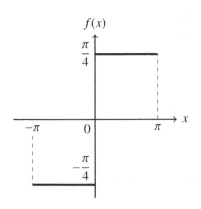

$$f(x)$$

Solution for b_n

$$b_n \pi = \text{sum of cases}$$

$$= -\frac{\pi}{4}(-\pi, 0) + \frac{\pi}{4}(0, \pi)$$

apply Rule 5b

$$= -\frac{\pi}{4}(\pi, 0) + \frac{\pi}{4}(0, \pi)$$

$$= \frac{\pi}{4}(0, \pi) + \frac{\pi}{4}(0, \pi)$$

$$= \frac{\pi}{2}(0, \pi)$$

apply Rule 3b

$$b_n \pi = \frac{\pi}{2} \cdot \frac{1}{n}(1 - \cos n\pi)$$

$$b_n = \frac{1}{2n}(1 - \cos n\pi)$$

Therefore

$$b_n = \begin{cases} \dfrac{1}{n} & n \text{ is odd} \\ 0 & n \text{ is even} \end{cases}$$

Solution for Fourier series

$$f(x) = \sum_{n=1}^{\infty} b_n \sin nx$$

$$= b_n(\sin x + \sin 3x + \cdots)$$
$$+ b_n(\sin 2x + \sin 4x + \cdots)$$

$$= \frac{1}{n}(\sin x + \sin 3x + \cdots)$$
$$+ 0(\sin 2x + \sin 4x + \cdots)$$

$$= (\sin x + \frac{1}{3}\sin 3x + \cdots) + 0$$

$$\boxed{f(x) = \sin x + \frac{1}{3}\sin 3x + \cdots}$$

3.

$$f(x) = \begin{cases} 2 & -2 < x < 0 \\ 0 & 0 < x < 2 \end{cases}$$

$$f(x + 4) = f(x)$$

Determine the property of $f(x)$ The graph of $f(x)$ below shows that it is neither an even nor odd function.

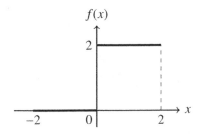

$$f(x)$$

Solution for ω The period T is given as
4.

$$\omega = \frac{2\pi}{T}$$

$$= \frac{2\pi}{4} = \frac{\pi}{2}$$

Solution for b_n

$$b_n \frac{\pi}{\omega} = \text{sum of cases}$$

put $\omega = \dfrac{\pi}{2}$

$$2b_n = 2(-2, 0) + 0(0, 2)$$

$$= 2(-2, 0)$$

$$= -2(0, -2) \qquad \text{(RP1)}$$

apply Rule 5b

$$2b_n = -2(0, 2)$$

$$b_n = -(0, 2)$$

$$= -\left[\frac{1}{n\omega}[1 - \cos n\omega(2)]\right]$$

$$= -\left[\frac{2}{n\pi}(1 - \cos n\pi)\right]$$

$$b_n = \frac{2}{n\pi}(\cos n\pi - 1)$$

Therefore

$$b_n = \begin{cases} -\dfrac{4}{n\pi} & n \text{ is odd} \\ 0 & n \text{ is even} \end{cases}$$

Solution for a_n Continuing from RP1,

$$2a_n = -2(0, -2)$$

apply Rule 5a

$$= 2(0, 2) \qquad\qquad \text{(RP2)}$$

$$= 2\frac{1}{n\omega}\sin n\omega(2)$$

$$= 2\frac{2}{n\pi}\sin n\pi$$

$$2a_n = 0$$

$$\boxed{a_n = 0}$$

Solution for a_0 Continuing from RP2,

$$2a_0 = 2(0, 2)$$

$$= 2(2)$$

$$\boxed{a_0 = 2}$$

Solution for Fourier series

$$f(x) = \frac{a_0}{2} + \sum_{n=1}^{\infty} a_n \cos n\omega x + \sum_{n=1}^{\infty} b_n \sin n\omega x$$

$$= \frac{2}{2} + 0(\cos\frac{\pi x}{2} + \cos 2\frac{\pi x}{2} + \cdots)$$

$$+ b_n(\sin\frac{\pi x}{2} + \sin 3\frac{\pi x}{2} + \cdots)$$

$$+ b_n(\sin 2\frac{\pi x}{2} + \sin 4\frac{\pi x}{2} + \cdots)$$

$$= 1 - \frac{4}{n\pi}(\sin\frac{\pi x}{2} + \sin 3\frac{\pi x}{2} + \cdots)$$

$$+ 0(\sin 2\frac{\pi x}{2} + \sin 4\frac{\pi x}{2} + \cdots)$$

$$\boxed{f(x) = 1 - \frac{4}{\pi}\left(\sin\frac{\pi x}{2} + \frac{1}{3}\sin 3\frac{\pi x}{2} + \cdots\right)}$$

4.

$$f(x) = \begin{cases} x & -\dfrac{\pi}{2} < x < \dfrac{\pi}{2} \\ \pi - x & \dfrac{\pi}{2} < x < \dfrac{3\pi}{2} \end{cases}$$

$$f(x + 2\pi) = f(x)$$

Determine the property of $f(x)$ The graph of $f(x)$ below shows that it is an odd function. Thus

$$\boxed{a_n = 0} \qquad \text{and} \qquad \boxed{a_0 = 0}$$

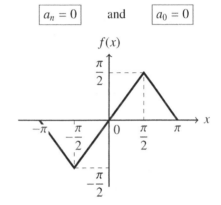

Solution for ω The period T is given as 2π.

$$\omega = \frac{2\pi}{T}$$

$$= \frac{2\pi}{2\pi} = 1$$

Solution for b_n

$$b_n\pi$$

$$= \text{sum of cases}$$

$$= x(-\frac{\pi}{2}, \frac{\pi}{2}) + (\pi - x)(\frac{\pi}{2}, \frac{3\pi}{2})$$

$$= x(-\frac{\pi}{2}, 0) + x(0, \frac{\pi}{2}) + \pi(\frac{\pi}{2}, \frac{3\pi}{2})$$

$$- x(\frac{\pi}{2}, \frac{3\pi}{2})$$

$$= -x(0, -\frac{\pi}{2}) + x(0, \frac{\pi}{2}) + \pi(\frac{\pi}{2}, 0) + \pi(0, \frac{3\pi}{2})$$

$$- x(\frac{\pi}{2}, 0) - x(0, \frac{3\pi}{2})$$

$$= -x(0, -\frac{\pi}{2}) + x(0, \frac{\pi}{2}) - \pi(0, \frac{\pi}{2}) + \pi(0, \frac{3\pi}{2})$$

$$+ x(0, \frac{\pi}{2}) - x(0, \frac{3\pi}{2})$$

$$= -x(0, -\frac{\pi}{2}) + 2x(0, \frac{\pi}{2}) - \pi(0, \frac{\pi}{2}) + \pi(0, \frac{3\pi}{2})$$

$$- x(0, \frac{3\pi}{2}) \qquad \text{(RP1)}$$

apply Rule 6b

$$= x(0, \frac{\pi}{2}) + 2x(0, \frac{\pi}{2}) - \pi(0, \frac{\pi}{2}) + \pi(0, \frac{3\pi}{2})$$

$$- x(0, \frac{3\pi}{2})$$

$$= 3x(0, \frac{\pi}{2}) - \pi(0, \frac{\pi}{2}) + \pi(0, \frac{3\pi}{2}) - x(0, \frac{3\pi}{2})$$

$$= 3\left[\frac{-\pi/2}{n} \cos \frac{n\pi}{2} + \frac{1}{n^2} \sin \frac{n\pi}{2}\right]$$

$$- \frac{\pi}{n}\left(1 - \cos \frac{n\pi}{2}\right) + \frac{\pi}{n}\left(1 - \cos \frac{3n\pi}{2}\right)$$

$$- \left[\frac{-3\pi/2}{n} \cos \frac{3n\pi}{2} + \frac{1}{n^2} \sin \frac{3n\pi}{2}\right]$$

$$= -\frac{3\pi}{2n} \cos \frac{n\pi}{2} + \frac{3}{n^2} \sin \frac{n\pi}{2}$$

$$+ \frac{\pi}{n}\left(-1 + \cos \frac{n\pi}{2} + 1 - \cos \frac{3n\pi}{2}\right)$$

$$+ \frac{3\pi}{2n} \cos \frac{3n\pi}{2} - \frac{1}{n^2} \sin \frac{3n\pi}{2}$$

$$= \frac{3\pi}{2n}\left(\cos \frac{3n\pi}{2} - \cos \frac{n\pi}{2}\right)$$

$$+ \frac{1}{n^2}\left(3 \sin \frac{n\pi}{2} - \sin \frac{3n\pi}{2}\right)$$

$$+ \frac{\pi}{n}\left(\cos \frac{n\pi}{2} - \cos \frac{3n\pi}{2}\right)$$

We can simplify the result using

$$\cos \frac{3n\pi}{2} = \cos(n\pi + \frac{n\pi}{2})$$

$$= \cos n\pi \cos \frac{n\pi}{2} + \sin n\pi \sin \frac{n\pi}{2}$$

$$= \cos n\pi \cos \frac{n\pi}{2}$$

$$= \cos \frac{n\pi}{2}$$

and

$$\sin \frac{3n\pi}{2} = \sin(n\pi + \frac{n\pi}{2})$$

$$= \sin n\pi \cos \frac{n\pi}{2} + \cos n\pi \sin \frac{n\pi}{2}$$

$$= \cos n\pi \sin \frac{n\pi}{2}$$

$$= -\sin \frac{n\pi}{2}$$

Therefore

$$b_n\pi = \frac{3\pi}{2n}\left(\cos \frac{n\pi}{2} - \cos \frac{n\pi}{2}\right)$$

$$+ \frac{1}{n^2}\left(3 \sin \frac{n\pi}{2} + \sin \frac{n\pi}{2}\right)$$

$$+ \frac{\pi}{n}\left(\cos \frac{n\pi}{2} - \cos \frac{n\pi}{2}\right)$$

$$= \frac{4}{n^2} \sin \frac{n\pi}{2}$$

$$b_n = \frac{4}{\pi n^2} \sin \frac{n\pi}{2}$$

Recall that

$$\sin \frac{n\pi}{2} = \begin{cases} 0 & n \text{ is even} \\ 1 & n = 1, 5, 9, \ldots \\ -1 & n = 3, 7, 11, \ldots \end{cases}$$

Therefore

$$b_n = \begin{cases} 0 & n \text{ is even} \\ \dfrac{4}{\pi n^2} & n = 1, 5, 9, \ldots \\ -\dfrac{4}{\pi n^2} & n = 3, 7, 11, \ldots \end{cases}$$

Solution for Fourier series

$$f(x) = \sum_{n=1}^{\infty} b_n \sin nx$$

$$= 0\left(\sin 2x + \sin 4x + \cdots\right)$$

$$+ b_n\left(\sin x + \sin 5x + \cdots\right)$$

$$+ b_n\left(\sin 3x + \sin 7x + \cdots\right)$$

$$= \frac{4}{\pi n^2}\left(\sin x + \sin 5x + \cdots\right)$$

$$- \frac{4}{\pi n^2}\left(\sin 3x + \sin 7x + \cdots\right)$$

$$= \frac{4}{\pi n^2}\left(\sin x - \sin 3x + \cdots\right)$$

$$\boxed{f(x) = \frac{4}{\pi}\left(\sin x - \frac{1}{3^2}\sin 3x + \cdots\right)}$$

5.

$$f(x) = \begin{cases} 1 - \frac{1}{2}|x| & -2 < x < 2 \\ 0 & 2 < x < 6 \end{cases}$$

$$f(x + 8) = f(x)$$

Solution for analytical definition

$$|x| = \begin{cases} -x & x < 0 \\ x & x \geq 0 \end{cases}$$

$$-\frac{1}{2}|x| = \begin{cases} \dfrac{x}{2} & x < 0 \\ -\dfrac{x}{2} & x \geq 0 \end{cases}$$

$$1 - \frac{1}{2}|x| = \begin{cases} 1 + \dfrac{x}{2} & x < 0 \\ 1 - \dfrac{x}{2} & x \geq 0 \end{cases}$$

Now

$$f(x) = \begin{cases} 1 - \frac{1}{2}|x| & -2 < x < 0 \\ 1 - \frac{1}{2}|x| & 0 < x < 2 \\ 0 & 2 < x < 6 \end{cases}$$

Therefore

$$\boxed{f(x) = \begin{cases} 1 + \dfrac{x}{2} & -2 < x < 0 \\ 1 - \dfrac{x}{2} & 0 < x < 2 \\ 0 & 2 < x < 6 \end{cases}}$$

Determine the property of $f(x)$ The graph of $f(x)$ below shows that it is an even function. Thus

$$\boxed{b_n = 0}$$

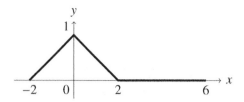

Solution for ω The period T is given as 8.

$$\omega = \frac{2\pi}{T}$$

$$= \frac{2\pi}{8} = \frac{\pi}{4}$$

Solution for a_n

$$a_n \frac{\pi}{\omega} = \text{sum of cases}$$

$$\text{put } \omega = \frac{\pi}{4}$$

$$4a_n = (1 + \frac{x}{2})(-2,0)$$

$$+ (1 - \frac{x}{2})(0,2) + 0(2,6)$$

$$= (-2,0) + \frac{x}{2}(-2,0)$$

$$+ (0,2) - \frac{x}{2}(0,2)$$

$$= -(0,-2) - \frac{x}{2}(0,-2)$$

$$+ (0,2) - \frac{x}{2}(0,2)$$

apply Rules 5a and 6a

$$= (0,2) - \frac{x}{2}(0,2)$$

$$+ (0,2) - \frac{x}{2}(0,2)$$

$$= 2(0,2) - x(0,2) \tag{RP}$$

$$= 2 \cdot \frac{1}{n\omega}\sin n\omega(2) - \left[\frac{2}{n\omega}\sin n\omega(2)\right.$$

$$+ \frac{1}{n^2\omega^2}(\cos n\omega(2) - 1)\Big]$$

$$= 2 \cdot \frac{4}{n\pi}\sin \frac{n\pi}{2}$$

$$- \left[\frac{8}{n\pi}\sin \frac{n\pi}{2} + \frac{16}{n^2\pi^2}(\cos \frac{n\pi}{2} - 1)\right]$$

$$= \frac{8}{n\pi} \sin \frac{n\pi}{2}$$

$$- \frac{8}{n\pi} \sin \frac{n\pi}{2} - \frac{16}{n^2\pi^2}(\cos \frac{n\pi}{2} - 1)$$

$$4a_n = 0 - \frac{16}{n^2\pi^2}(\cos \frac{n\pi}{2} - 1)$$

$$a_n = -\frac{4}{n^2\pi^2}(\cos \frac{n\pi}{2} - 1)$$

Recall that

$$\cos \frac{n\pi}{2} = \begin{cases} 0 & n \text{ is odd} \\ -1 & n = 2, 6, 10, \ldots \\ 1 & n = 4, 8, 12, \ldots \end{cases}$$

Therefore

$$a_n = \begin{cases} \dfrac{4}{n^2\pi^2} & n \text{ is odd} \\ \dfrac{8}{n^2\pi^2} & n = 2, 6, 10, \ldots \\ 0 & n = 4, 8, 12, \ldots \end{cases}$$

Solution for a_0   Continuing from RP,

$$4a_0 = 2(0, 2) - x(0, 2)$$

$$= 2(2) - \frac{2^2}{2}$$

$$= 2$$

$$\boxed{a_0 = \frac{1}{2}}$$

Solution for Fourier series

$$f(x)$$

$$= \frac{a_0}{2} + \sum_{n=1}^{\infty} a_n \cos n\omega x$$

$$= \frac{1}{4} + a_n(\cos \frac{\pi x}{4} + \cos 3\frac{\pi x}{4} + \cdots)$$

$$+ a_n(\cos 2\frac{\pi x}{4} + \cos 6\frac{\pi x}{4} + \cdots)$$

$$+ 0(\cos 4\frac{\pi x}{4} + \cos 8\frac{\pi x}{4} + \cdots)$$

$$= \frac{1}{4} + \frac{4}{n^2\pi^2}(\cos \frac{\pi x}{4} + \cos 3\frac{\pi x}{4} + \cdots)$$

$$+ \frac{8}{n^2\pi^2}(\cos 2\frac{\pi x}{4} + \cos 6\frac{\pi x}{4} + \cdots)$$

$$\boxed{\begin{aligned} f(x) &= \frac{1}{4} + \frac{4}{\pi^2}(\cos \frac{\pi x}{4} + \frac{1}{3^2} \cos 3\frac{\pi x}{4} + \cdots) \\ &+ \frac{8}{\pi^2}(\frac{1}{2^2} \cos 2\frac{\pi x}{4} + \frac{1}{6^2} \cos 6\frac{\pi x}{4} + \cdots) \end{aligned}}$$

Chapter 7

1.

$$g(x) = 1 \qquad 0 < x < 2$$

Solution for ω   The period T of $f(x)$ will be twice the upper bound of $g(x)$. That is,

$$T = 2 \times 2 = 4$$

Thus

$$\omega = \frac{2\pi}{T}$$

$$= \frac{2\pi}{4}$$

$$\boxed{\omega = \frac{\pi}{2}}$$

Solution for b_n

$$b_n \frac{\pi}{\omega} = 2 \times \text{sum of cases}$$

$$\text{put } \omega = \frac{\pi}{2}$$

$$2b_n = 2 \times 1(0, 2)$$

$$b_n = (0, 2) \qquad\qquad \text{(RP)}$$

$$= \frac{1}{n\omega}[1 - \cos n\omega(2)]$$

$$= \frac{2}{n\pi}(1 - \cos n\pi)$$

$$\boxed{b_n = \begin{cases} \dfrac{4}{n\pi} & n \text{ is odd} \\ 0 & n \text{ is even} \end{cases}}$$

Solution for a_n   Continuing from RP,

$a_n = (0, 2)$

$= \dfrac{1}{n\omega} \sin n\omega(2)$

$= \dfrac{2}{n\pi} \sin n\pi$

$\boxed{a_n = 0}$

Solution for a_0   Continuing from RP,

$a_0 = (0, 2)$

$\boxed{a_0 = 2}$

Solution for sine series

$f(x) = \displaystyle\sum_{n=1}^{\infty} b_n \sin n\omega x$

$= b_n \left(\sin \dfrac{\pi x}{2} + \sin 3\dfrac{\pi x}{2} + \cdots \right)$

$\quad + b_n \left(\sin 2\dfrac{\pi x}{2} + \sin 4\dfrac{\pi x}{2} + \cdots \right)$

$= \dfrac{4}{n\pi} \left(\sin \dfrac{\pi x}{2} + \sin 3\dfrac{\pi x}{2} + \cdots \right)$

$\quad + 0 \left(\sin 2\dfrac{\pi x}{2} + \sin 4\dfrac{\pi x}{2} + \cdots \right)$

$\boxed{f(x) = \dfrac{4}{\pi} \left(\sin \dfrac{\pi x}{2} + \dfrac{1}{3} \sin 3\dfrac{\pi x}{2} + \cdots \right)}$

Solution for cosine series

$f(x) = \dfrac{a_0}{2} + \displaystyle\sum_{n=1}^{\infty} a_n \cos n\omega x$

$= \dfrac{2}{2} + 0 \left(\cos \dfrac{\pi x}{2} + \cos 3\dfrac{\pi x}{2} + \cdots \right)$

$\boxed{f(x) = 1}$

2.

$g(x) = x \qquad 0 < x < \dfrac{1}{2}$

Solution for ω   The period T of $f(x)$ will be twice the upper bound of $g(x)$. That is,

$$T = 2 \times \dfrac{1}{2} = 1$$

Thus

$$\omega = \dfrac{2\pi}{T}$$

$$= \dfrac{2\pi}{1}$$

$$\boxed{\omega = 2\pi}$$

Solution for b_n

$b_n \dfrac{\pi}{\omega} = 2 \times \text{sum of cases}$

$\quad$ put $\omega = 2\pi$

$\dfrac{1}{2} b_n = 2 \times x(0, \dfrac{1}{2})$

$b_n = 4x(0, \dfrac{1}{2})$ \hfill (RP)

$= 4 \left[\dfrac{-1/2}{n\omega} \cos n\omega(\dfrac{1}{2}) + \dfrac{1}{n^2\omega^2} \sin n\omega(\dfrac{1}{2}) \right]$

$= -\dfrac{1}{n\pi} \cos n\pi + \dfrac{4}{4n^2\pi^2} \sin n\pi$

$= -\dfrac{1}{n\pi} \cos n\pi$

$\boxed{b_n = \begin{cases} \dfrac{1}{n\pi} & n \text{ is odd} \\[2mm] -\dfrac{1}{n\pi} & n \text{ is even} \end{cases}}$

Solution for a_n   Continuing from RP,

$a_n = 4x(0, \dfrac{1}{2})$

$= 4 \left[\dfrac{1/2}{n\omega} \sin n\omega(\dfrac{1}{2}) \right.$

$\quad \left. + \dfrac{1}{n^2\omega^2} \left(\cos n\omega(\dfrac{1}{2}) - 1 \right) \right]$

$= \dfrac{1}{n\pi} \sin n\pi + \dfrac{4}{4n^2\pi^2} (\cos n\pi - 1)$

$= \dfrac{1}{n^2\pi^2} (\cos n\pi - 1)$

$$a_n = \begin{cases} -\dfrac{2}{n^2\pi^2} & n \text{ is odd} \\ 0 & n \text{ is even} \end{cases}$$

Solution for a_0 Continuing from RP,

$$a_0 = 4x(0, \tfrac{1}{2})$$

$$= 4 \times \frac{(1/2)^2}{2}$$

$$\boxed{a_0 = \frac{1}{2}}$$

Solution for sine series

$$f(x) = \sum_{n=1}^{\infty} b_n \sin n\omega x$$

$$= b_n(\sin(2\pi x) + \sin 3(2\pi x) + \cdots)$$
$$+ b_n(\sin 2(2\pi x) + \sin 4(2\pi x) + \cdots)$$

$$= \frac{1}{n\pi}(\sin(2\pi x) + \sin 3(2\pi x) + \cdots)$$
$$- \frac{1}{n\pi}(\sin 2(2\pi x) + \sin 4(2\pi x) + \cdots)$$

$$= \frac{1}{n\pi}(\sin(2\pi x) - \sin 2(2\pi x) + \cdots)$$

$$\boxed{f(x) = \frac{1}{\pi}(\sin(2\pi x) - \frac{1}{2}\sin 2(2\pi x) + \cdots)}$$

Solution for cosine series

$$f(x) = \frac{a_0}{2} + \sum_{n=1}^{\infty} a_n \cos n\omega x$$

$$= \frac{1}{4} + a_n(\cos(2\pi x) + \cos 3(2\pi x) + \cdots)$$
$$+ a_n(\cos 2(2\pi x) + \cos 4(2\pi x) + \cdots)$$

$$= \frac{1}{4} - \frac{2}{n^2\pi^2}(\cos(2\pi x)$$
$$+ \cos 3(2\pi x) + \cdots)$$
$$+ 0(\cos 2(2\pi x) + \cos 4(2\pi x) + \cdots)$$

$$\boxed{\begin{aligned} &f(x) \\ &= \frac{1}{4} - \frac{2}{\pi^2}(\cos(2\pi x) + \frac{1}{3^2} \cos 3(2\pi x) + \cdots) \end{aligned}}$$

3.

$$g(x) = 4 \qquad 0 < x < 3$$

Solution for ω The period T of $f(x)$ will be twice the upper bound of $g(x)$. That is,

$$T = 2 \times 3 = 6$$

Thus

$$\omega = \frac{2\pi}{T}$$

$$= \frac{2\pi}{6}$$

$$\boxed{\omega = \frac{\pi}{3}}$$

Solution for b_n

$$b_n \frac{\pi}{\omega} = 2 \times \text{sum of cases}$$

$$\text{put } \omega = \frac{\pi}{3}$$

$$3b_n = 2 \times 4(0, 3)$$

$$= 8(0, 3) \qquad\qquad \text{(RP)}$$

$$= 8 \cdot \frac{1}{n\omega}[1 - \cos n\omega(3)]$$

$$3b_n = \frac{24}{n\pi}(1 - \cos n\pi)$$

$$b_n = \frac{8}{n\pi}(1 - \cos n\pi)$$

$$\boxed{b_n = \begin{cases} \dfrac{16}{n\pi} & n \text{ is odd} \\ 0 & n \text{ is even} \end{cases}}$$

Solution for a_n   Continuing from RP,

$$3a_n = 8(0, 3)$$

$$= 8 \cdot \frac{1}{n\omega} \sin n\omega(3)$$

$$= \frac{24}{n\pi} \sin n\pi$$

$$3a_n = 0$$

$$\boxed{a_n = 0}$$

Solution for a_0   Continuing from RP,

$$3a_0 = 8(0, 3)$$

$$= 8(3)$$

$$\boxed{a_0 = 8}$$

Solution for sine series

$$f(x) = \sum_{n=1}^{\infty} b_n \sin n\omega x$$

$$= b_n\left(\sin \frac{\pi x}{3} + \sin 3\frac{\pi x}{3} + \cdots \right)$$

$$+ b_n\left(\sin 2\frac{\pi x}{3} + \sin 4\frac{\pi x}{3} + \cdots \right)$$

$$= \frac{16}{n\pi}\left(\sin \frac{\pi x}{3} + \sin 3\frac{\pi x}{3} + \cdots \right)$$

$$+ 0\left(\sin \frac{\pi x}{3} + \sin 2\frac{\pi x}{3} + \cdots \right)$$

$$\boxed{f(x) = \frac{16}{\pi}\left(\sin \frac{\pi x}{3} + \frac{1}{3} \sin 3\frac{\pi x}{3} + \cdots \right)}$$

Solution for cosine series

$$f(x) = \frac{a_0}{2} + \sum_{n=1}^{\infty} a_n \cos n\omega x$$

$$= \frac{8}{2} + 0\left(\cos \frac{\pi x}{3} + \cos 3\frac{\pi x}{3} + \cdots \right)$$

$$\boxed{f(x) = 4}$$

4.

$$g(x) = \begin{cases} 0 & 0 < x < 2 \\ 1 & 2 < x < 4 \end{cases}$$

Solution for ω   The period T of $f(x)$ will be twice the upper bound of $g(x)$. That is,

$$T = 2 \times 4 = 8$$

Thus

$$\omega = \frac{2\pi}{T}$$

$$= \frac{2\pi}{8}$$

$$\boxed{\omega = \frac{\pi}{4}}$$

Solution for b_n

$$b_n \frac{\pi}{\omega} = 2 \times \text{sum of cases}$$

$$\text{put } \omega = \frac{\pi}{4}$$

$$4b_n = 2\left[0(0, 2) + 1(2, 4)\right]$$

$$= 2(2, 4)$$

$$= 2(2, 0) + 2(0, 4)$$

$$= -2(0, 2) + 2(0, 4) \qquad \text{(RP)}$$

$$= -2\frac{1}{n\omega}[1 - \cos n\omega(2)]$$

$$+ 2\frac{1}{n\omega}[1 - \cos n\omega(4)]$$

$$4b_n = -\frac{8}{n\pi}\left(1 - \cos \frac{n\pi}{2}\right) + \frac{8}{n\pi}(1 - \cos n\pi)$$

$$b_n = -\frac{2}{n\pi}\left(1 - \cos \frac{n\pi}{2}\right) + \frac{2}{n\pi}(1 - \cos n\pi)$$

$$= \frac{2}{n\pi}\left(-1 + \cos \frac{n\pi}{2} + 1 - \cos n\pi\right)$$

$$= \frac{2}{n\pi}\left(\cos \frac{n\pi}{2} - \cos n\pi\right)$$

Recall that

$$\cos n\pi = \begin{cases} -1 & n \text{ is odd} \\ 1 & n \text{ is even} \end{cases}$$

and

$$\cos \frac{n\pi}{2} = \begin{cases} 0 & n \text{ is odd} \\ -1 & n = 2, 6, 10, \ldots \\ 1 & n = 4, 8, 12, \ldots \end{cases}$$

When n is odd

$$b_n = \frac{2}{n\pi}[0 - (-1)]$$

$$= \frac{2}{n\pi}$$

When $n = 2, 6, 10, \ldots$

$$b_n = \frac{2}{n\pi}(-1 - 1)$$

$$= -\frac{4}{n\pi}$$

When $n = 4, 8, 12, \ldots$

$$b_n = \frac{2}{n\pi}(1 - 1)$$

$$= 0$$

Therefore

$$b_n = \begin{cases} \dfrac{2}{n\pi} & n \text{ is odd} \\ -\dfrac{4}{n\pi} & n = 2, 6, 10, \ldots \\ 0 & n = 4, 8, 12, \ldots \end{cases}$$

Solution for a_n   Continuing from RP,

$$4a_n = -2(0, 2) + 2(0, 4)$$

$$= -2\frac{1}{n\omega}\sin n\omega(2) + 2\frac{1}{n\omega}\sin n\omega(4)$$

$$4a_n = -\frac{8}{n\pi}\sin\frac{n\pi}{2} + \frac{8}{n\pi}\sin n\pi$$

$$a_n = -\frac{2}{n\pi}\sin\frac{n\pi}{2}$$

Recall that

$$\sin\frac{n\pi}{2} = \begin{cases} 0 & n \text{ is even} \\ 1 & n = 1, 5, 9, \ldots \\ -1 & n = 3, 7, 11, \ldots \end{cases}$$

Therefore

$$a_n = \begin{cases} 0 & n \text{ is even} \\ -\dfrac{2}{n\pi} & n = 1, 5, 9, \ldots \\ \dfrac{2}{n\pi} & n = 3, 7, 11, \ldots \end{cases}$$

Solution for a_0   Continuing from RP,

$$4a_0 = -2(0, 2) + 2(0, 4)$$

$$= -2(2) + 2(4)$$

$$= -4 + 8$$

$$4a_0 = 4$$

$$\boxed{a_0 = 1}$$

Solution for sine series

$$f(x) = \sum_{n=1}^{\infty} b_n \sin n\omega x$$

$$= b_n\left(\sin\frac{\pi x}{4} + \sin 3\frac{\pi x}{4} + \cdots\right)$$

$$+ b_n\left(\sin 2\frac{\pi x}{4} + \sin 6\frac{\pi x}{4} + \cdots\right)$$

$$+ b_n\left(\sin 4\frac{\pi x}{4} + \sin 8\frac{\pi x}{4} + \cdots\right)$$

$$= \frac{2}{n\pi}\left(\sin\frac{\pi x}{4} + \sin 3\frac{\pi x}{4} + \cdots\right)$$

$$- \frac{4}{n\pi}\left(\sin 2\frac{\pi x}{4} + \sin 6\frac{\pi x}{4} + \cdots\right)$$

$$+ 0\left(\sin 4\frac{\pi x}{4} + \sin 8\frac{\pi x}{4} + \cdots\right)$$

$$\boxed{\begin{aligned} f(x) = & \frac{2}{\pi}\left(\sin\frac{\pi x}{4} + \frac{1}{3}\sin 3\frac{\pi x}{4} + \cdots\right) \\ & - \frac{4}{\pi}\left(\frac{1}{2}\sin 2\frac{\pi x}{4} + \frac{1}{6}\sin 6\frac{\pi x}{4} + \cdots\right) \end{aligned}}$$

Solution for cosine series

$$f(x) = \frac{a_0}{2} + \sum_{n=1}^{\infty} a_n \cos n\omega x$$

$$= \frac{1}{2} + a_n\left(\cos 2\frac{\pi x}{4} + \cos 4\frac{\pi x}{4} + \cdots\right)$$

$$+ a_n\left(\cos\frac{\pi x}{4} + \cos 5\frac{\pi x}{4} + \cdots\right)$$

$$+ a_n\left(\cos 3\frac{\pi x}{4} + \cos 7\frac{\pi x}{4} + \cdots\right)$$

$$= \frac{1}{2} + 0(\cos 2\frac{\pi x}{4} + \cos 4\frac{\pi x}{4} + \cdots)$$

$$- \frac{2}{n\pi}(\cos\frac{\pi x}{4} + \cos 5\frac{\pi x}{4} + \cdots)$$

$$+ \frac{2}{n\pi}(\cos 3\frac{\pi x}{4} + \cos 7\frac{\pi x}{4} + \cdots)$$

$$= \frac{1}{2} - \frac{2}{n\pi}(\cos\frac{\pi x}{4} - \cos 3\frac{\pi x}{4} + \cdots)$$

$$\boxed{f(x) = \frac{1}{2} - \frac{2}{\pi}(\cos\frac{\pi x}{4} - \frac{1}{3}\cos 3\frac{\pi x}{4} + \cdots)}$$

5.

$$g(x) = \begin{cases} 1 & 0 < x < 1 \\ 2 & 1 < x < 2 \end{cases}$$

Solution for ω The period T of $f(x)$ will be twice the upper bound of $g(x)$. That is,

$$T = 2 \times 2 = 4$$

Thus

$$\omega = \frac{2\pi}{T}$$

$$= \frac{2\pi}{4}$$

$$\boxed{\omega = \frac{\pi}{2}}$$

Solution for b_n

$$b_n\frac{\pi}{\omega} = 2 \times \text{sum of cases}$$

$$\text{put } \omega = \frac{\pi}{2}$$

$$2b_n = 2\Big[1(0,1) + 2(1,2)\Big]$$

$$= 2(0,1) + 4(1,2)$$

$$b_n = (0,1) + 2(1,2)$$

$$= (0,1) + 2(1,0) + 2(0,2)$$

$$= (0,1) - 2(0,1) + 2(0,2)$$

$$= -(0,1) + 2(0,2) \qquad \text{(RP)}$$

$$= -\frac{1}{n\omega}[1 - \cos n\omega(1)]$$

$$+ 2\frac{1}{n\omega}[1 - \cos n\omega(2)]$$

$$= -\frac{2}{n\pi}(1 - \cos\frac{n\pi}{2})$$

$$+ \frac{4}{n\pi}(1 - \cos n\pi)$$

$$= \frac{2}{n\pi}(-1 + \cos\frac{n\pi}{2} + 2 - 2\cos n\pi)$$

$$= \frac{2}{n\pi}(\cos\frac{n\pi}{2} - 2\cos n\pi + 1)$$

Recall that

$$\cos n\pi = \begin{cases} -1 & n \text{ is odd} \\ 1 & n \text{ is even} \end{cases}$$

and

$$\cos\frac{n\pi}{2} = \begin{cases} 0 & n \text{ is odd} \\ -1 & n = 2, 6, 10, \ldots \\ 1 & n = 4, 8, 12, \ldots \end{cases}$$

When n **is odd**

$$b_n = \frac{2}{n\pi}(0 - 2(-1) + 1)$$

$$= \frac{6}{n\pi}$$

When $n = 2, 6, 10, \ldots$

$$b_n = \frac{2}{n\pi}(-1 - 2(1) + 1)$$

$$= -\frac{4}{n\pi}$$

When $n = 4, 8, 12, \ldots$

$$b_n = \frac{2}{n\pi}(1 - 2(1) + 1)$$

$$= 0$$

Therefore

$$b_n = \begin{cases} \dfrac{6}{n\pi} & n \text{ is odd} \\ -\dfrac{4}{n\pi} & n = 2, 6, 10, \ldots \\ 0 & n = 4, 8, 12, \ldots \end{cases}$$

Solution for a_n   Continuing from RP,

$$a_n = -(0, 1) + 2(0, 2)$$

$$= -\frac{1}{n\omega} \sin n\omega(1) + 2\frac{1}{n\omega} \sin n\omega(2)$$

$$= -\frac{2}{n\pi} \sin \frac{n\pi}{2} + \frac{4}{n\pi} \sin n\pi$$

$$a_n = -\frac{2}{n\pi} \sin \frac{n\pi}{2}$$

Recall that

$$\sin \frac{n\pi}{2} = \begin{cases} 0 & n \text{ is even} \\ 1 & n = 1, 5, 9, \ldots \\ -1 & n = 3, 7, 11, \ldots \end{cases}$$

Therefore

$$a_n = \begin{cases} 0 & n \text{ is even} \\ -\dfrac{2}{n\pi} & n = 1, 5, 9, \ldots \\ \dfrac{2}{n\pi} & n = 3, 7, 11, \ldots \end{cases}$$

Solution for a_0   Continuing from RP,

$$a_0 = -(0, 1) + 2(0, 2)$$

$$= -1 + 2(2)$$

$$\boxed{a_0 = 3}$$

Solution for sine series

$$f(x) = \sum_{n=1}^{\infty} b_n \sin n\omega x$$

$$= b_n \left(\sin \frac{\pi x}{2} + \sin 3\frac{\pi x}{2} + \cdots \right)$$

$$+ b_n \left(\sin 2\frac{\pi x}{2} + \sin 6\frac{\pi x}{2} + \cdots \right)$$

$$+ b_n \left(\sin 4\frac{\pi x}{2} + \sin 8\frac{\pi x}{2} + \cdots \right)$$

$$= \frac{6}{n\pi} \left(\sin \frac{\pi x}{2} + \sin 3\frac{\pi x}{2} + \cdots \right)$$

$$- \frac{4}{n\pi} \left(\sin 2\frac{\pi x}{2} + \sin 6\frac{\pi x}{2} + \cdots \right)$$

$$+ 0 \left(\sin 4\frac{\pi x}{2} + \sin 8\frac{\pi x}{2} + \cdots \right)$$

$$\boxed{f(x) = \frac{6}{\pi} \left(\sin \frac{\pi x}{2} + \frac{1}{3} \sin 3\frac{\pi x}{2} + \cdots \right) \\ - \frac{4}{\pi} \left(\frac{1}{2} \sin 2\frac{\pi x}{2} + \frac{1}{6} \sin 6\frac{\pi x}{2} + \cdots \right)}$$

Solution for cosine series

$$f(x) = \frac{a_0}{2} + \sum_{n=1}^{\infty} a_n \cos n\omega x$$

$$= \frac{3}{2} + a_n \left(\cos 2\frac{\pi x}{2} + \cos 4\frac{\pi x}{2} + \cdots \right)$$

$$+ a_n \left(\cos \frac{\pi x}{2} + \cos 5\frac{\pi x}{2} + \cdots \right)$$

$$+ a_n \left(\cos 3\frac{\pi x}{2} + \cos 7\frac{\pi x}{2} + \cdots \right)$$

$$= \frac{3}{2} + 0 \left(\cos 2\frac{\pi x}{2} + \cos 4\frac{\pi x}{2} + \cdots \right)$$

$$- \frac{2}{n\pi} \left(\cos \frac{\pi x}{2} + \cos 5\frac{\pi x}{2} + \cdots \right)$$

$$+ \frac{2}{n\pi} \left(\cos 3\frac{\pi x}{2} + \cos 7\frac{\pi x}{2} + \cdots \right)$$

$$= \frac{3}{2} - \frac{2}{n\pi} \left(\cos \frac{\pi x}{2} - \cos 3\frac{\pi x}{2} + \cdots \right)$$

$$\boxed{f(x) = \frac{3}{2} - \frac{2}{\pi} \left(\cos \frac{\pi x}{2} - \frac{1}{3} \cos 3\frac{\pi x}{2} + \cdots \right)}$$

6.

$$g(x) = \begin{cases} 0 & 0 < x < \dfrac{\pi}{2} \\ 1 & \dfrac{\pi}{2} < x < \pi \end{cases}$$

Solution for ω   The period T of $f(x)$ will be twice the upper bound of $g(x)$. That is,

$$T = 2 \times \pi = 2\pi$$

Thus

$$\omega = \frac{2\pi}{T}$$

$$= \frac{2\pi}{2\pi}$$

$$\boxed{\omega = 1}$$

Solution for b_n

$$b_n \frac{\pi}{\omega} = 2 \times \text{sum of cases}$$

$$\text{put } \omega = 1$$

$$b_n \pi = 2 \left[0(0, \frac{\pi}{2}) + 1(\frac{\pi}{2}, \pi) \right]$$

$$= 2(\frac{\pi}{2}, \pi)$$

$$= 2(\frac{\pi}{2}, 0) + 2(0, \pi)$$

$$= -2(0, \frac{\pi}{2}) + 2(0, \pi) \qquad \text{(RP)}$$

$$= -2 \frac{1}{n\omega} [1 - \cos n\omega(\frac{\pi}{2})]$$

$$+ 2 \frac{1}{n\omega} [1 - \cos n\omega(\pi)]$$

$$= -\frac{2}{n} (1 - \cos \frac{n\pi}{2}) + \frac{2}{n} (1 - \cos n\pi)$$

$$b_n \pi = \frac{2}{n} (-1 + \cos \frac{n\pi}{2} + 1 - \cos n\pi)$$

$$b_n = \frac{2}{n\pi} (\cos \frac{n\pi}{2} - \cos n\pi)$$

Recall that

$$\cos n\pi = \begin{cases} -1 & n \text{ is odd} \\ 1 & n \text{ is even} \end{cases}$$

and

$$\cos \frac{n\pi}{2} = \begin{cases} 0 & n \text{ is odd} \\ -1 & n = 2, 6, 10, \dots \\ 1 & n = 4, 8, 12, \dots \end{cases}$$

When n is odd

$$b_n = \frac{2}{n\pi} [0 - (-1)]$$

$$= \frac{2}{n\pi}$$

When $n = 2, 6, 10, \dots$

$$b_n = \frac{2}{n\pi} (-1 - 1)$$

$$= -\frac{4}{n\pi}$$

When $n = 4, 8, 12, \dots$

$$b_n = \frac{2}{n\pi} (1 - 1)$$

$$= 0$$

Therefore

$$b_n = \begin{cases} \dfrac{2}{n\pi} & n \text{ is odd} \\ -\dfrac{4}{n\pi} & n = 2, 6, 10, \dots \\ 0 & n = 4, 8, 12, \dots \end{cases}$$

Solution for a_n   Continuing from RP,

$$a_n \pi = -2(0, \frac{\pi}{2}) + 2(0, \pi)$$

$$= -2 \frac{1}{n\omega} \sin n\omega(\frac{\pi}{2})$$

$$+ 2 \frac{1}{n\omega} \sin n\omega(\pi)$$

$$a_n \pi = -\frac{2}{n} \sin \frac{n\pi}{2} + \frac{2}{n} \sin n\pi$$

$$a_n = -\frac{2}{n\pi} \sin \frac{n\pi}{2}$$

Recall that

$$\sin \frac{n\pi}{2} = \begin{cases} 0 & n \text{ is even} \\ 1 & n = 1, 5, 9, \dots \\ -1 & n = 3, 7, 11, \dots \end{cases}$$

Therefore

$$a_n = \begin{cases} 0 & n \text{ is even} \\ -\dfrac{2}{n\pi} & n = 1, 5, 9, \dots \\ \dfrac{2}{n\pi} & n = 3, 7, 11, \dots \end{cases}$$

Solution for a_0   Continuing from RP,

$$a_0\pi = -2(0, \frac{\pi}{2}) + 2(0, \pi)$$

$$= -2(\frac{\pi}{2}) + 2(\pi)$$

$$= \pi$$

$$\boxed{a_0 = 1}$$

Solution for sine series

$$f(x) = \sum_{n=1}^{\infty} b_n \sin n\omega x$$

$$= b_n(\sin x + \sin 3x + \cdots)$$
$$+ b_n(\sin 2x + \sin 6x + \cdots)$$
$$+ b_n(\sin 4x + \sin 8x + \cdots)$$

$$= \frac{2}{n\pi}(\sin x + \sin 3x + \cdots)$$

$$- \frac{4}{n\pi}(2 \sin x + \sin 6x + \cdots)$$

$$+ 0(\sin 2x + \sin 8x + \cdots)$$

$$\boxed{\begin{aligned} f(x) &= \frac{2}{\pi}(\sin x + \frac{1}{3}\sin 3x + \cdots) \\ &- \frac{4}{\pi}(\frac{1}{2}\sin 2x + \frac{1}{6}\sin 6x + \cdots) \end{aligned}}$$

Solution for cosine series

$$f(x) = \frac{a_0}{2} + \sum_{n=1}^{\infty} a_n \cos n\omega x$$

$$= \frac{1}{2} + a_n(\cos 2x + \cos 4x + \cdots)$$
$$+ a_n(\cos x + \cos 5x + \cdots)$$
$$+ a_n(\cos 3x + \cos 7x + \cdots)$$

$$= \frac{1}{2} + 0(\cos 2x + \cos 4x + \cdots)$$

$$- \frac{2}{n\pi}(\cos x + \cos 5x + \cdots)$$

$$+ \frac{2}{n\pi}(\cos 3x + \cos 7x + \cdots)$$

$$= \frac{1}{2} - \frac{2}{n\pi}(\cos x - \cos 3x + \cdots)$$

$$\boxed{f(x) = \frac{1}{2} - \frac{2}{\pi}(\cos x - \frac{1}{3}\cos 3x + \cdots)}$$

7.

$$g(x) = \begin{cases} x & 0 < x < \frac{\pi}{2} \\ \pi - x & \frac{\pi}{2} < x < \pi \end{cases}$$

Solution for ω   The period T of $f(x)$ will be twice the upper bound of $g(x)$. That is,

$$T = 2 \times \pi = 2\pi$$

Thus

$$\omega = \frac{2\pi}{T}$$

$$= \frac{2\pi}{2\pi}$$

$$\boxed{\omega = 1}$$

Solution for b_n

$$b_n \frac{\pi}{\omega} = 2 \times \text{sum of cases}$$

put $\omega = 1$

$$b_n \pi = 2\left[x(0, \frac{\pi}{2}) + (\pi - x)(\frac{\pi}{2}, \pi) \right]$$

$$\frac{1}{2}b_n \pi = x(0, \frac{\pi}{2}) + (\pi - x)(\frac{\pi}{2}, \pi)$$

$$= x(0, \frac{\pi}{2}) + \pi(\frac{\pi}{2}, \pi) - x(\frac{\pi}{2}, \pi)$$

$$= x(0, \frac{\pi}{2}) + \pi(\frac{\pi}{2}, 0) + \pi(0, \pi)$$

$$\quad - x(\frac{\pi}{2}, 0) - x(0, \pi)$$

$$= x(0, \frac{\pi}{2}) - \pi(0, \frac{\pi}{2}) + \pi(0, \pi)$$

$$\quad + x(0, \frac{\pi}{2}) - x(0, \pi)$$

$$= 2x(0, \frac{\pi}{2}) - \pi(0, \frac{\pi}{2})$$

$$\quad + \pi(0, \pi) - x(0, \pi) \qquad \text{(RP)}$$

$$= 2\left[-\frac{\pi/2}{n}\cos\frac{n\pi}{2} + \frac{1}{n^2}\sin\frac{n\pi}{2} \right]$$

$$\quad - \pi\left[\frac{1}{n}(1 - \cos\frac{n\pi}{2}) \right]$$

$$\quad + \pi\left[\frac{1}{n}(1 - \cos n\pi) \right]$$

$$\quad - \left[-\frac{\pi}{n}\cos n\pi + \frac{1}{n^2}\sin n\pi \right]$$

$$= -\frac{\pi}{n}\cos\frac{n\pi}{2} + \frac{2}{n^2}\sin\frac{n\pi}{2}$$

$$- \frac{\pi}{n}(1 - \cos\frac{n\pi}{2})$$

$$+ \frac{\pi}{n}(1 - \cos n\pi) + \frac{\pi}{n}\cos n\pi$$

$$= \frac{\pi}{n}(-\cos\frac{n\pi}{2} - 1 + \cos\frac{n\pi}{2} + 1$$

$$- \cos n\pi + \cos n\pi) + \frac{2}{n^2}\sin\frac{n\pi}{2}$$

$$\frac{1}{2}b_n\pi = \frac{\pi}{n}(0) + \frac{2}{n^2}\sin\frac{n\pi}{2}$$

$$b_n = \frac{4}{\pi n^2}\sin\frac{n\pi}{2}$$

Recall that

$$\sin\frac{n\pi}{2} = \begin{cases} 0 & n \text{ is even} \\ 1 & n = 1, 5, 9, \ldots \\ -1 & n = 3, 7, 11, \ldots \end{cases}$$

Therefore

$$b_n = \begin{cases} 0 & n \text{ is even} \\ \dfrac{4}{\pi n^2} & n = 1, 5, 9, \ldots \\ -\dfrac{4}{\pi n^2} & n = 3, 7, 11, \ldots \end{cases}$$

Solution for a_n   Continuing from RP,

$$\frac{1}{2}a_n\pi = 2x(0, \frac{\pi}{2}) - \pi(0, \frac{\pi}{2})$$

$$+ \pi(0, \pi) - x(0, \pi)$$

$$= 2\left[\frac{\pi/2}{n}\sin\frac{n\pi}{2} + \frac{1}{n^2}(\cos\frac{n\pi}{2} - 1)\right]$$

$$- \pi \cdot \frac{1}{n}\sin\frac{n\pi}{2} + \pi \cdot \frac{1}{n}\sin n\pi$$

$$- \left[\frac{\pi}{n}\sin n\pi + \frac{1}{n^2}(\cos n\pi - 1)\right]$$

$$= \frac{\pi}{n}\sin\frac{n\pi}{2} + \frac{2}{n^2}(\cos\frac{n\pi}{2} - 1)$$

$$- \frac{\pi}{n}\sin\frac{n\pi}{2} - \frac{1}{n^2}(\cos n\pi - 1)$$

$$= \frac{2}{n^2}(\cos\frac{n\pi}{2} - 1) - \frac{1}{n^2}(\cos n\pi - 1)$$

$$\frac{1}{2}a_n\pi = \frac{1}{n^2}(2\cos\frac{n\pi}{2} - 2 - \cos n\pi + 1)$$

$$a_n = \frac{2}{\pi n^2}(2\cos\frac{n\pi}{2} - \cos n\pi - 1)$$

Recall that

$$\cos n\pi = \begin{cases} -1 & n \text{ is odd} \\ 1 & n \text{ is even} \end{cases}$$

and

$$\cos\frac{n\pi}{2} = \begin{cases} 0 & n \text{ is odd} \\ -1 & n = 2, 6, 10, \ldots \\ 1 & n = 4, 8, 12, \ldots \end{cases}$$

When n is odd

$$a_n = \frac{2}{\pi n^2}(2(0) - (-1) - 1)$$

$$= 0$$

When $n = 2, 6, 10, \ldots$

$$a_n = \frac{2}{\pi n^2}(2(-1) - 1 - 1)$$

$$= -\frac{8}{\pi n^2}$$

When $n = 4, 8, 12, \ldots$

$$a_n = \frac{2}{\pi n^2}(2(1) - 1 - 1)$$

$$= 0$$

Therefore

$$a_n = \begin{cases} 0 & n \text{ is odd} \\ -\dfrac{8}{\pi n^2} & n = 2, 6, 10, \ldots \\ 0 & n = 4, 8, 12, \ldots \end{cases}$$

Solution for a_0   Continuing from RP,

$$\frac{1}{2}a_0\pi = 2x(0, \frac{\pi}{2}) - \pi(0, \frac{\pi}{2})$$

$$+ \pi(0, \pi) - x(0, \pi)$$

$$= 2 \cdot \frac{(\pi/2)^2}{2} - \pi(\frac{\pi}{2}) + \pi(\pi) - \frac{\pi^2}{2}$$

$$= \frac{\pi^2}{4} - \frac{\pi^2}{2} + \pi^2 - \frac{\pi^2}{2}$$

$$\frac{1}{2}a_0\pi = \frac{\pi^2}{4}$$

$$\boxed{a_0 = \frac{\pi}{2}}$$

Solution for sine series

$$f(x) = \sum_{n=1}^{\infty} b_n \sin n\omega x$$

$$= b_n(\sin 2x + \sin 4x + \cdots)$$

$$+ b_n(\sin x + \sin 5x + \cdots)$$

$$+ b_n(\sin 3x + \sin 7x + \cdots)$$

$$= 0(\sin 2x + \sin 4x + \cdots)$$

$$+ \frac{4}{\pi n^2}(\sin x + \sin 5x + \cdots)$$

$$- \frac{4}{\pi n^2}(\sin 3x + \sin 7x + \cdots)$$

$$= \frac{4}{\pi n^2}(\sin x - \sin 3x + \cdots)$$

$$\boxed{f(x) = \frac{4}{\pi}(\sin x - \frac{1}{3^2}\sin 3x + \cdots)}$$

Solution for cosine series

$$f(x) = \frac{a_0}{2} + \sum_{n=1}^{\infty} a_n \cos n\omega x$$

$$= \frac{\pi}{4} + a_n(\cos x + \cos 3x + \cdots)$$

$$+ a_n(\cos 2x + \cos 6x + \cdots)$$

$$+ a_n(\cos 4x + \cos 8x + \cdots)$$

$$= \frac{\pi}{4} + 0(\cos x + \cos 3x + \cdots)$$

$$- \frac{8}{\pi n^2}(\cos 2x + \cos 6x + \cdots)$$

$$+ 0(\cos 4x + \cos 8x + \cdots)$$

$$\boxed{\begin{array}{l} f(x) \\ = \dfrac{\pi}{4} - \dfrac{8}{\pi}(\dfrac{1}{2^2}\cos 2x + \dfrac{1}{6^2}\cos 6x + \cdots) \end{array}}$$

8.

$$g(x) = \begin{cases} \dfrac{1}{2} & 0 < x < \dfrac{5}{2} \\ 1 & \dfrac{5}{2} < x < 5 \end{cases}$$

Solution for ω   The period T of $f(x)$ will be twice the upper bound of $g(x)$. That is,

$$T = 2 \times 5 = 10$$

Thus

$$\omega = \frac{2\pi}{T}$$

$$= \frac{2\pi}{10}$$

$$\boxed{\omega = \frac{\pi}{5}}$$

Solution for b_n

$$b_n \frac{\pi}{\omega} = 2 \times \text{sum of cases}$$

$$\text{put } \omega = \frac{\pi}{5}$$

$$5b_n = 2\left[\frac{1}{2}(0, \frac{5}{2}) + 1(\frac{5}{2}, 5)\right]$$

$$= (0, \frac{5}{2}) + 2(\frac{5}{2}, 5)$$

$$= (0, \frac{5}{2}) + 2(\frac{5}{2}, 0) + 2(0, 5)$$

$$= (0, \frac{5}{2}) - 2(0, \frac{5}{2}) + 2(0, 5)$$

$$= -(0, \frac{5}{2}) + 2(0, 5) \qquad \text{(RP)}$$

$$= -\frac{1}{n\omega}[1 - \cos n\omega(\frac{5}{2})]$$

$$+ 2\frac{1}{n\omega}[1 - \cos n\omega(5)]$$

$$= -\frac{5}{n\pi}(1 - \cos \frac{n\pi}{2}) + \frac{10}{n\pi}(1 - \cos n\pi)$$

$$5b_n = \frac{5}{n\pi}\left(-1 + \cos\frac{n\pi}{2} + 2 - 2\cos n\pi\right)$$

$$b_n = \frac{1}{n\pi}\left(\cos\frac{n\pi}{2} - 2\cos n\pi + 1\right)$$

Recall that

$$\cos n\pi = \begin{cases} -1 & n \text{ is odd} \\ 1 & n \text{ is even} \end{cases}$$

and

$$\cos\frac{n\pi}{2} = \begin{cases} 0 & n \text{ is odd} \\ -1 & n = 2, 6, 10, \ldots \\ 1 & n = 4, 8, 12, \ldots \end{cases}$$

When n is odd

$$b_n = \frac{1}{n\pi}(0 - 2(-1) + 1)$$

$$= \frac{3}{n\pi}$$

When $n = 2, 6, 10, \ldots$

$$b_n = \frac{1}{n\pi}(-1 - 2(1) + 1)$$

$$= -\frac{2}{n\pi}$$

When $n = 4, 8, 12, \ldots$

$$b_n = \frac{1}{n\pi}(1 - 2(1) + 1)$$

$$= 0$$

Therefore

$$b_n = \begin{cases} \dfrac{3}{n\pi} & n \text{ is odd} \\[2mm] -\dfrac{2}{n\pi} & n = 2, 6, 10, \ldots \\[2mm] 0 & n = 4, 8, 12, \ldots \end{cases}$$

Solution for a_n   Continuing from RP,

$$5a_n = -\left(0, \frac{5}{2}\right) + 2(0, 5)$$

$$= -\frac{1}{n\omega}\sin n\omega\left(\frac{5}{2}\right)$$

$$+ 2\frac{1}{n\omega}\sin n\omega(5)$$

$$= -\frac{5}{n\pi}\sin\frac{n\pi}{2} + 2\frac{5}{n\pi}\sin n\pi$$

$$5a_n = -\frac{5}{n\pi}\sin\frac{n\pi}{2}$$

$$a_n = -\frac{1}{n\pi}\sin\frac{n\pi}{2}$$

Recall that

$$\sin\frac{n\pi}{2} = \begin{cases} 0 & n \text{ is even} \\ 1 & n = 1, 5, 9, \ldots \\ -1 & n = 3, 7, 11, \ldots \end{cases}$$

Therefore

$$a_n = \begin{cases} 0 & n \text{ is even} \\[2mm] -\dfrac{1}{n\pi} & n = 1, 5, 9, \ldots \\[2mm] \dfrac{1}{n\pi} & n = 3, 7, 11, \ldots \end{cases}$$

Solution for a_0   Continuing from RP,

$$5a_n = -\left(0, \frac{5}{2}\right) + 2(0, 5)$$

$$= -\frac{5}{2} + 2(5)$$

$$a_0 = -\frac{1}{2} + 2$$

$$\boxed{a_0 = \frac{3}{2}}$$

Solution for sine series

$$f(x) = \sum_{n=1}^{\infty} b_n \sin n\omega x$$

$$= b_n\left(\sin\frac{\pi x}{5} + \sin 3\frac{\pi x}{5} + \cdots\right)$$

$$+ b_n\left(\sin 2\frac{\pi x}{5} + \sin 6\frac{\pi x}{5} + \cdots\right)$$

$$+ b_n\left(\sin 4\frac{\pi x}{5} + \sin 8\frac{\pi x}{5} + \cdots\right)$$

$$= \frac{3}{n\pi}\left(\sin\frac{\pi x}{5} + \sin 3\frac{\pi x}{5} + \cdots\right)$$

$$-\frac{2}{n\pi}(\sin 2\frac{\pi x}{5} + \sin 6\frac{\pi x}{5} + \cdots)$$

$$+ 0(\sin 4\frac{\pi x}{5} + \sin 8\frac{\pi x}{5} + \cdots)$$

$$\boxed{\begin{aligned} f(x) &= \frac{3}{\pi}(\sin\frac{\pi x}{5} + \frac{1}{3}\sin 3\frac{\pi x}{5} + \cdots) \\ &\quad - \frac{2}{\pi}(\frac{1}{2}\sin 2\frac{\pi x}{5} + \frac{1}{6}\sin 6\frac{\pi x}{5} + \cdots) \end{aligned}}$$

Solution for cosine series

$$f(x) = \frac{a_0}{2} + \sum_{n=1}^{\infty} a_n \cos n\omega x$$

$$= \frac{3}{4} + a_n(\cos 2\frac{\pi x}{5} + \cos 4\frac{\pi x}{5} + \cdots)$$

$$+ a_n(\cos\frac{\pi x}{5} + \cos 5\frac{\pi x}{5} + \cdots)$$

$$+ a_n(\cos 3\frac{\pi x}{5} + \cos 7\frac{\pi x}{5} + \cdots)$$

$$= \frac{3}{2} + 0(\cos 2\frac{\pi x}{5} + \cos 4\frac{\pi x}{5} + \cdots)$$

$$- \frac{1}{n\pi}(\cos\frac{\pi x}{5} + \cos 5\frac{\pi x}{5} + \cdots)$$

$$+ \frac{1}{n\pi}(\cos 3\frac{\pi x}{5} + \cos 7\frac{\pi x}{5} + \cdots)$$

$$= \frac{3}{4} - \frac{1}{n\pi}(\cos\frac{\pi x}{5} - \cos 3\frac{\pi x}{5} + \cdots)$$

$$\boxed{f(x) = \frac{3}{4} - \frac{1}{\pi}(\cos\frac{\pi x}{5} - \frac{1}{3}\cos 3\frac{\pi x}{5} + \cdots)}$$

9.

$$g(x) = \begin{cases} x & 0 < x < \frac{1}{2} \\ 1-x & \frac{1}{2} < x < 1 \end{cases}$$

Solution for ω The period T of $f(x)$ will be twice the upper bound of $g(x)$. That is,

$$T = 2 \times 1 = 2$$

Thus

$$\omega = \frac{2\pi}{T}$$

$$= \frac{2\pi}{2}$$

$$\boxed{\omega = \pi}$$

Solution for b_n

$$b_n\frac{\pi}{\omega} = 2 \times \text{sum of cases}$$

put $\omega = \pi$

$$b_n = 2\left[x(0, \frac{1}{2}) + (1-x)(\frac{1}{2}, 1)\right]$$

$$\frac{1}{2}b_n = x(0, \frac{1}{2}) + (1-x)(\frac{1}{2}, 1)$$

$$= x(0, \frac{1}{2}) + (\frac{1}{2}, 1) - x(\frac{1}{2}, 1)$$

$$= x(0, \frac{1}{2}) + (\frac{1}{2}, 0) + (0, 1)$$

$$- x(\frac{1}{2}, 0) - x(0, 1)$$

$$= x(0, \frac{1}{2}) - (0, \frac{1}{2}) + (0, 1)$$

$$+ x(0, \frac{1}{2}) - x(0, 1)$$

$$= 2x(0, \frac{1}{2}) - (0, \frac{1}{2})$$

$$+ (0, 1) - x(0, 1) \qquad \text{(RP)}$$

$$= 2\left[\frac{-1/2}{n\omega}\cos n\omega(\frac{1}{2}) + \frac{1}{n^2\omega^2}\sin n\omega(\frac{1}{2})\right]$$

$$- \frac{1}{n\omega}[1 - \cos n\omega(\frac{1}{2})]$$

$$+ \frac{1}{n\omega}[1 - \cos n\omega(1)]$$

$$- \left[\frac{-1}{n\omega}\cos n\omega(1) + \frac{1}{n^2\omega^2}\sin n\omega(1)\right]$$

$$= -\frac{1}{n\pi}\cos\frac{n\pi}{2} + \frac{2}{n^2\pi^2}\sin\frac{n\pi}{2}$$

$$- \frac{1}{n\pi}(1 - \cos\frac{n\pi}{2}) + \frac{1}{n\pi}(1 - \cos n\pi)$$

$$+ \frac{1}{n\pi}\cos n\pi - \frac{1}{n^2\pi^2}\sin n\pi$$

$$\frac{1}{2}b_n = \frac{1}{n\pi}(-\cos\frac{n\pi}{2} - 1 + \cos\frac{n\pi}{2} + 1$$

$$- \cos n\pi + \cos n\pi) + \frac{2}{n^2\pi^2}\sin\frac{n\pi}{2}$$

$$b_n = \frac{4}{n^2\pi^2}\sin\frac{n\pi}{2}$$

Recall that

$$\sin\frac{n\pi}{2} = \begin{cases} 0 & n \text{ is even} \\ 1 & n = 1, 5, 9, \ldots \\ -1 & n = 3, 7, 11, \ldots \end{cases}$$

Therefore

$$b_n = \begin{cases} 0 & n \text{ is even} \\ \dfrac{4}{n^2\pi^2} & n = 1, 5, 9, \ldots \\ -\dfrac{4}{n^2\pi^2} & n = 3, 7, 11, \ldots \end{cases}$$

Solution for a_n   Continuing from RP,

$$\frac{1}{2}a_n$$

$$= 2x(0, \frac{1}{2}) - (0, \frac{1}{2}) + (0, 1) - x(0, 1)$$

$$= 2\left[\frac{1/2}{n\omega}\sin n\omega(\frac{1}{2}) + \frac{1}{n^2\omega^2}(\cos n\omega(\frac{1}{2}) - 1)\right]$$

$$\quad - \frac{1}{n\omega}\sin n\omega(\frac{1}{2}) + \frac{1}{n\omega}\sin n\omega(1)$$

$$\quad - \left[\frac{1}{n\omega}\sin n\omega(1) + \frac{1}{n^2\omega^2}(\cos n\omega(1) - 1)\right]$$

$$= \frac{1}{n\pi}\sin\frac{n\pi}{2} + \frac{2}{n^2\pi^2}(\cos\frac{n\pi}{2} - 1)$$

$$\quad - \frac{1}{n\pi}\sin\frac{n\pi}{2} + \frac{1}{n\pi}\sin n\pi$$

$$\quad - \frac{1}{n\pi}\sin n\pi - \frac{1}{n^2\pi^2}(\cos n\pi - 1)$$

$$= \frac{2}{n^2\pi^2}(\cos\frac{n\pi}{2} - 1) - \frac{1}{n^2\pi^2}(\cos n\pi - 1)$$

$$= \frac{1}{n^2\pi^2}(2\cos\frac{n\pi}{2} - 2 - \cos n\pi + 1)$$

$$= \frac{1}{n^2\pi^2}(2\cos\frac{n\pi}{2} - \cos n\pi - 1)$$

Therefore

$$a_n = \frac{2}{n^2\pi^2}(2\cos\frac{n\pi}{2} - \cos n\pi - 1)$$

Recall that

$$\cos n\pi = \begin{cases} -1 & n \text{ is odd} \\ 1 & n \text{ is even} \end{cases}$$

and

$$\cos\frac{n\pi}{2} = \begin{cases} 0 & n \text{ is odd} \\ -1 & n = 2, 6, 10, \ldots \\ 1 & n = 4, 8, 12, \ldots \end{cases}$$

When n is odd

$$a_n = \frac{2}{n^2\pi^2}(2(0) - (-1) - 1)$$

$$= 0$$

When $n = 2, 6, 10, \ldots$

$$a_n = \frac{2}{n^2\pi^2}(2(-1) - (1) - 1)$$

$$= -\frac{8}{n^2\pi^2}$$

When $n = 4, 8, 12, \ldots$

$$a_n = \frac{2}{n^2\pi^2}(2(1) - (1) - 1)$$

$$= 0$$

Therefore

$$a_n = \begin{cases} 0 & n \text{ is odd} \\ -\dfrac{8}{n^2\pi^2} & n = 2, 6, 10, \ldots \\ 0 & n = 4, 8, 12, \ldots \end{cases}$$

Solution for a_0   Continuing from RP,

$$\frac{1}{2}a_0 = 2x(0, \frac{1}{2}) - (0, \frac{1}{2}) + (0, 1) - x(0, 1)$$

$$= 2 \cdot \frac{(1/2)^2}{2} - \frac{1}{2} + 1 - \frac{1^2}{2}$$

$$= \frac{1}{4} - \frac{1}{2} + 1 - \frac{1}{2}$$

$$a_0 = \frac{1}{2}$$

Solution for sine series

$$f(x) = \sum_{n=1}^{\infty} b_n \sin n\omega x$$

$$= b_n(\sin 2\pi x + \sin 4\pi x + \cdots)$$
$$+ b_n(\sin \pi x + \sin 5\pi x + \cdots)$$
$$+ b_n(\sin 3\pi x + \sin 7\pi x + \cdots)$$
$$= 0(\sin 2\pi x + \sin 4\pi x + \cdots)$$
$$+ \frac{4}{n^2\pi^2}(\sin \pi x + \sin 5\pi x + \cdots)$$
$$- \frac{4}{n^2\pi^2}(\sin 3\pi x + \sin 7\pi x + \cdots)$$
$$= \frac{4}{n^2\pi^2}(\sin \pi x - \sin 3\pi x + \cdots)$$

$$\boxed{f(x) = \frac{4}{\pi^2}(\sin \pi x - \frac{1}{3^2}\sin 3\pi x + \cdots)}$$

Solution for cosine series

$$f(x) = \frac{a_0}{2} + \sum_{n=1}^{\infty} a_n \cos n\omega x$$

$$= \frac{1}{4} + a_n(\cos \pi x + \cos 3\pi x + \cdots)$$
$$+ a_n(\cos 2\pi x + \cos 6\pi x + \cdots)$$
$$+ a_n(\cos 4\pi x + \cos 8\pi x + \cdots)$$
$$= \frac{1}{4} + 0(\cos \pi x + \cos 3\pi x + \cdots)$$
$$- \frac{8}{n^2\pi^2}(\cos 2\pi x + \cos 6\pi x + \cdots)$$
$$+ 0(\cos 4\pi x + \cos 8\pi x + \cdots)$$

$$\boxed{\begin{array}{l} f(x) \\ = \frac{1}{4} - \frac{8}{\pi^2}(\frac{1}{2^2}\cos 2\pi x + \frac{1}{6^2}\cos 6\pi x + \cdots) \end{array}}$$

10.

$$g(x) = \begin{cases} 1 & 0 < x < \dfrac{5}{2} \\ -1 & \dfrac{5}{2} < x < 5 \end{cases}$$

Solution for ω   The period T of $f(x)$ will be twice the upper bound of $g(x)$. That is,

$$T = 2 \times 5 = 10$$

Thus

$$\omega = \frac{2\pi}{T}$$
$$= \frac{2\pi}{10}$$

$$\boxed{\omega = \frac{\pi}{5}}$$

Solution for b_n

$$b_n\frac{\pi}{\omega} = 2 \times \text{sum of cases}$$

$$\text{put } \omega = \frac{\pi}{5}$$

$$5b_n = 2\left[1(0, \frac{5}{2}) - 1(\frac{5}{2}, 5)\right]$$

$$= 2(0, \frac{5}{2}) - 2(\frac{5}{2}, 5)$$

$$= 2(0, \frac{5}{2}) - 2(\frac{5}{2}, 0) - 2(0, 5)$$

$$= 2(0, \frac{5}{2}) + 2(0, \frac{5}{2}) - 2(0, 5)$$

$$= 4(0, \frac{5}{2}) - 2(0, 5) \qquad \text{(RP)}$$

$$= 4\frac{1}{n\omega}[1 - \cos n\omega(\frac{5}{2})]$$

$$- 2\frac{1}{n\omega}[1 - \cos n\omega(5)]$$

$$= \frac{20}{n\pi}(1 - \cos \frac{n\pi}{2}) - \frac{10}{n\pi}(1 - \cos n\pi)$$

$$5b_n = \frac{10}{n\pi}(2 - 2\cos \frac{n\pi}{2} - 1 + \cos n\pi)$$

$$b_n = \frac{2}{n\pi}(\cos n\pi - 2\cos \frac{n\pi}{2} + 1)$$

Recall that

$$\cos n\pi = \begin{cases} -1 & n \text{ is odd} \\ 1 & n \text{ is even} \end{cases}$$

and

$$\cos \frac{n\pi}{2} = \begin{cases} 0 & n \text{ is odd} \\ -1 & n = 2, 6, 10, \ldots \\ 1 & n = 4, 8, 12, \ldots \end{cases}$$

When n is odd

$$b_n = \frac{2}{n\pi}(-1 - 2(0) + 1)$$
$$= 0$$

When $n = 2, 6, 10, \ldots$

$$b_n = \frac{2}{n\pi}(1 - 2(-1) + 1)$$
$$= \frac{8}{n\pi}$$

When $n = 4, 8, 12, \ldots$

$$b_n = \frac{2}{n\pi}(1 - 2(1) + 1)$$
$$= 0$$

Therefore

$$b_n = \begin{cases} 0 & n \text{ is odd} \\ \dfrac{8}{n\pi} & n = 2, 6, 10, \ldots \\ 0 & n = 4, 8, 12, \ldots \end{cases}$$

Solution for a_n Continuing from RP,

$$5a_n = 4(0, \frac{5}{2}) - 2(0, 5)$$
$$= 4\frac{1}{n\omega}\sin n\omega(\frac{5}{2}) - 2\frac{1}{n\omega}\sin n\omega(5)$$
$$= \frac{20}{n\pi}\sin \frac{n\pi}{2} - \frac{10}{n\pi}\sin n\pi$$
$$a_n = \frac{4}{n\pi}\sin \frac{n\pi}{2}$$

Recall that

$$\sin \frac{n\pi}{2} = \begin{cases} 0 & n \text{ is even} \\ 1 & n = 1, 5, 9, \ldots \\ -1 & n = 3, 7, 11, \ldots \end{cases}$$

Therefore

$$a_n = \begin{cases} 0 & n \text{ is even} \\ \dfrac{4}{n\pi} & n = 1, 5, 9, \ldots \\ -\dfrac{4}{n\pi} & n = 3, 7, 11, \ldots \end{cases}$$

Solution for a_0 Continuing from RP,

$$5a_n = 4(0, \frac{5}{2}) - 2(0, 5)$$
$$= 4(\frac{5}{2}) - 2(5)$$
$$a_0 = \frac{4}{2} - 2$$

$$\boxed{a_0 = 0}$$

Solution for sine series

$$f(x) = \sum_{n=1}^{\infty} b_n \sin n\omega x$$
$$= b_n(\sin \frac{\pi x}{5} + \sin 3\frac{\pi x}{5} + \cdots)$$
$$+ b_n(\sin 2\frac{\pi x}{5} + \sin 6\frac{\pi x}{5} + \cdots)$$
$$+ b_n(\sin 4\frac{\pi x}{5} + \sin 8\frac{\pi x}{5} + \cdots)$$
$$= 0(\sin \frac{\pi x}{5} + \sin 3\frac{\pi x}{5} + \cdots)$$
$$+ \frac{8}{n\pi}(\sin 2\frac{\pi x}{5} + \sin 6\frac{\pi x}{5} + \cdots)$$
$$+ 0(\sin 4\frac{\pi x}{5} + \sin 8\frac{\pi x}{5} + \cdots)$$

$$\boxed{f(x) = \frac{8}{\pi}(\frac{1}{2}\sin 2\frac{\pi x}{5} + \frac{1}{6}\sin 6\frac{\pi x}{5} + \cdots)}$$

Solution for cosine series

$$f(x) = \frac{a_0}{2} + \sum_{n=1}^{\infty} a_n \cos n\omega x$$

$$= \frac{0}{2} + a_n\left(\cos 2\frac{\pi x}{5} + \cos 4\frac{\pi x}{5} + \cdots\right)$$

$$+ a_n\left(\cos \frac{\pi x}{5} + \cos 5\frac{\pi x}{5} + \cdots\right)$$

$$+ a_n\left(\cos 3\frac{\pi x}{5} + \cos 7\frac{\pi x}{5} + \cdots\right)$$

$$= 0\left(\cos 2\frac{\pi x}{5} + \cos 4\frac{\pi x}{5} + \cdots\right)$$

$$+ \frac{4}{n\pi}\left(\cos \frac{\pi x}{5} + \cos 5\frac{\pi x}{5} + \cdots\right)$$

$$- \frac{4}{n\pi}\left(\cos 3\frac{\pi x}{5} + \cos 7\frac{\pi x}{5} + \cdots\right)$$

$$= \frac{4}{n\pi}\left(\cos \frac{\pi x}{5} - \cos 3\frac{\pi x}{5} + \cdots\right)$$

$$\boxed{f(x) = \frac{4}{\pi}\left(\cos \frac{\pi x}{5} - \frac{1}{3}\cos 3\frac{\pi x}{5} + \cdots\right)}$$

Chapter 8

Exercise 1

1.

$$f(x) = \begin{cases} -6 & -\pi < x < 0 \\ 6 & 0 < x < \pi \end{cases}$$

$$f(x) = \frac{24}{\pi}\left(\sin x + \frac{1}{3}\sin 3x + \cdots\right)$$

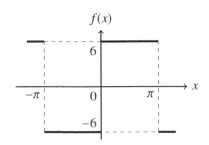

$f(x)$

Solution

(a) when $x = 0$

At $x = 0$, $f(x)$ is discontinuous.

$$\text{top value} = 6$$
$$\text{bottom value} = -6$$
$$\text{average} = \frac{6 + (-6)}{2}$$
$$= 0$$

Therefore, at $x = 0$, $f(x) = 0$. If we put $x = 0$, $f(x) = 0$ into the Fourier series, we get

$$0 = \frac{24}{\pi}\left(0 + \frac{1}{3}(0) + \cdots\right)$$
$$0 = 0$$

(b) when $x = \pi$

At $x = \pi$, $f(x)$ is discontinuous.

$$\text{top value} = 6$$
$$\text{bottom value} = -6$$
$$\text{average} = \frac{6 + (-6)}{2}$$
$$= 0$$

Therefore, at $x = \pi$, $f(x) = 0$. If we put $x = \pi$, $f(x) = 0$ into the Fourier series, we get

$$0 = \frac{24}{\pi}\left(0 + \frac{1}{3}(0) + \cdots\right)$$
$$0 = 0$$

(c) when $x = \frac{\pi}{2}$

At $x = \frac{\pi}{2}$, $f(x)$ is continuous and equal to 6.

If we put $x = \frac{\pi}{2}$, $f(x) = 6$ into the Fourier series, we get

$$6 = \frac{24}{\pi}\left(\sin\left(\frac{\pi}{2}\right) + \frac{1}{3}\sin 3\left(\frac{\pi}{2}\right) + \cdots\right)$$

$$\frac{6\pi}{24} = \sin\frac{\pi}{2} + \frac{1}{3}\sin\frac{3\pi}{2} + \cdots$$

Recall that

$$\sin \frac{n\pi}{2} = \begin{cases} 1 & n = 1, 5, 9, \ldots \\ -1 & n = 3, 7, 11, \ldots \end{cases}$$

Therefore

$$\frac{6\pi}{24} = 1 + \frac{1}{3}(-1) + \frac{1}{5}(1) + \frac{1}{7}(-1) + \cdots$$

$$\boxed{\frac{\pi}{4} = 1 - \frac{1}{3} + \frac{1}{5} - \frac{1}{7} + \cdots}$$

2.

$$f(x) = \begin{cases} 0 & -\pi < x < 0 \\ 6 & 0 < x < \pi \end{cases}$$

$$f(x) = 3 + \frac{12}{\pi}\left(\sin x + \frac{1}{3}\sin 3x + \cdots\right)$$

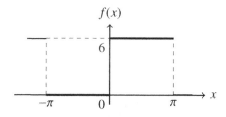

Solution

(a) when $x = 0$

At $x = 0$, $f(x)$ is discontinuous.

$$\text{top value} = 6$$
$$\text{bottom value} = 0$$
$$\text{average} = \frac{6 + 0}{2}$$
$$= 3$$

Therefore, at $x = 0$, $f(x) = 3$. If we put $x = 0$, $f(x) = 3$ into the Fourier series, we get

$$3 = 3 + \frac{12}{\pi}\left(0 + \frac{1}{3}(0) + \cdots\right)$$

$$0 = \frac{12}{\pi}\left(0 + 0 + \cdots\right)$$

$$0 = 0$$

(b) when $x = \pi$

At $x = \pi$, $f(x)$ is discontinuous.

$$\text{top value} = 6$$
$$\text{bottom value} = 0$$
$$\text{average} = \frac{6 + 0}{2}$$
$$= 3$$

Therefore, at $x = \pi$, $f(x) = 3$. If we put $x = \pi$, $f(x) = 3$ into the Fourier series, we get

$$3 = 3 + \frac{12}{\pi}\left(0 + \frac{1}{3}(0) + \cdots\right)$$

$$0 = \frac{12}{\pi}\left(0 + 0 + \cdots\right)$$

$$0 = 0$$

(c) when $x = \frac{\pi}{2}$

At $x = \frac{\pi}{2}$, $f(x)$ is continuous and equal to 6.

If we put $x = \frac{\pi}{2}$, $f(x) = 6$ into the Fourier series, we get

$$6 = 3 + \frac{12}{\pi}\left(\sin\left(\frac{\pi}{2}\right) + \frac{1}{3}\sin 3\left(\frac{\pi}{2}\right) + \cdots\right)$$

$$3 = \frac{12}{\pi}\left(\sin\left(\frac{\pi}{2}\right) + \frac{1}{3}\sin 3\left(\frac{\pi}{2}\right) + \cdots\right)$$

$$\frac{3\pi}{12} = \sin\frac{\pi}{2} + \frac{1}{3}\sin\frac{3\pi}{2} + \cdots$$

Recall that

$$\sin \frac{n\pi}{2} = \begin{cases} 1 & n = 1, 5, 9, \ldots \\ -1 & n = 3, 7, 11, \ldots \end{cases}$$

Therefore

$$\frac{3\pi}{12} = 1 + \frac{1}{3}(-1) + \frac{1}{5}(1) + \frac{1}{7}(-1) + \cdots$$

$$\boxed{\frac{\pi}{4} = 1 - \frac{1}{3} + \frac{1}{5} - \frac{1}{7} + \cdots}$$

3.

$$f(x) = x \qquad 0 < x < 2\pi$$

$$f(x) = \pi - 2\left(\sin x + \frac{1}{2}\sin 2x + \cdots\right)$$

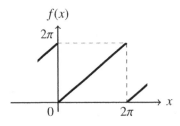

Solution

(a) when $x = 0$

At $x = 0$, $f(x)$ is discontinuous.

$$\text{top value} = 2\pi$$
$$\text{bottom value} = 0$$
$$\text{average} = \frac{2\pi + 0}{2}$$
$$= \pi$$

Therefore, at $x = 0$, $f(x) = \pi$. If we put $x = 0$, $f(x) = \pi$ into the Fourier series, we get

$$\pi = \pi - 2\left(\sin x + \frac{1}{2}\sin 2x + \cdots\right)$$
$$\pi = \pi - 2(0 + 0 + \cdots)$$
$$\pi = \pi$$

(b) when $x = \pi$

At $x = \pi$, $f(x)$ is continuous and equal to π.

If we put $x = \pi$, $f(x) = \pi$ into the Fourier series, we get

$$\pi = \pi - 2\left(\sin \pi + \frac{1}{2}\sin 2\pi + \cdots\right)$$
$$\pi = \pi - 2(0 + 0 + \cdots)$$
$$\pi = \pi$$

(c) when $x = \frac{\pi}{2}$

At $x = \frac{\pi}{2}$, $f(x)$ is continuous and equal to $\frac{\pi}{2}$.

If we put $x = \frac{\pi}{2}$, $f(x) = \frac{\pi}{2}$ into the Fourier series, we get

$$\frac{\pi}{2} = \pi - 2\left(\sin \frac{\pi}{2} + \frac{1}{2}\sin 2\frac{\pi}{2} + \cdots\right)$$
$$\frac{\pi}{2} = 2\left(\sin \frac{\pi}{2} + \frac{1}{2}\sin 2\frac{\pi}{2} + \cdots\right)$$

Recall that

$$\sin \frac{n\pi}{2} = \begin{cases} 0 & n \text{ is even} \\ 1 & n = 1, 5, 9, \ldots \\ -1 & n = 3, 7, 11, \ldots \end{cases}$$

Therefore

$$\frac{\pi}{2} = 2\left(1 + \frac{1}{2}(0) + \frac{1}{3}(-1) + \cdots\right)$$

$$\boxed{\frac{\pi}{4} = 1 - \frac{1}{3} + \frac{1}{5} - \frac{1}{7} + \cdots}$$

4.

$$f(x) = \begin{cases} x & -\frac{\pi}{2} < x < \frac{\pi}{2} \\ \pi - x & \frac{\pi}{2} < x < \frac{3\pi}{2} \end{cases}$$

$$f(x) = \frac{4}{\pi}\left(\sin x - \frac{1}{3^2}\sin 3x + \cdots\right)$$

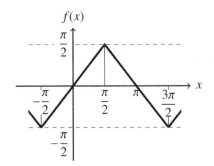

Solution

(a) when $x = 0$

At $x = 0$, $f(x)$ is continuous and equal to 0.

If we put $x = 0$, $f(x) = 0$ into the Fourier series, we get

$$0 = \frac{4}{\pi}\left(\sin 0 - \frac{1}{3^2}\sin 0 + \cdots\right)$$

$$0 = \frac{4}{\pi}\left(0 + 0 + \cdots\right)$$

$$0 = 0$$

(b) when $x = \pi$

At $x = \pi$, $f(x)$ is continuous and equal to 0.

If we put $x = \pi$, $f(x) = 0$ into the Fourier series, we get

$$0 = \frac{4}{\pi}\left(\sin \pi - \frac{1}{3^2}\sin 3\pi + \cdots\right)$$

$$0 = \frac{4}{\pi}\left(0 + 0 + \cdots\right)$$

$$0 = 0$$

(c) when $x = \dfrac{\pi}{2}$

At $x = \dfrac{\pi}{2}$, $f(x)$ is continuous and equal to $\dfrac{\pi}{2}$.

If we put $x = \dfrac{\pi}{2}$, $f(x) = \dfrac{\pi}{2}$ into the Fourier series, we get

$$\frac{\pi}{2} = \frac{4}{\pi}\left(\sin\frac{\pi}{2} - \frac{1}{3^2}\sin 3\frac{\pi}{2} + \cdots\right)$$

$$\frac{\pi^2}{8} = \sin\frac{\pi}{2} - \frac{1}{3^2}\sin 3\frac{\pi}{2} + \cdots$$

Recall that

$$\sin\frac{n\pi}{2} = \begin{cases} 1 & n = 1, 5, 9, \ldots \\ -1 & n = 3, 7, 11, \ldots \end{cases}$$

Therefore

$$\frac{\pi^2}{8} = 1 - \frac{1}{3^2}(-1) + \frac{1}{5^2}(1) - \frac{1}{7^2}(-1) + \cdots$$

$$\boxed{\frac{\pi^2}{8} = 1 + \frac{1}{3^2} + \frac{1}{5^2} + \frac{1}{7^2} + \cdots}$$

5.

$$f(x) = \begin{cases} \pi + x & -\pi < x < -\dfrac{\pi}{2} \\ \dfrac{\pi}{2} & -\dfrac{\pi}{2} < x < \dfrac{\pi}{2} \\ \pi - x & \dfrac{\pi}{2} < x < \pi \end{cases}$$

$$f(x) = \frac{3\pi}{8} + \frac{2}{\pi}\left(\cos x + \frac{1}{3^2}\cos 3x + \cdots\right)$$
$$- \frac{4}{\pi}\left(\frac{1}{2^2}\cos 2x + \frac{1}{6^2}\cos 6x + \cdots\right)$$

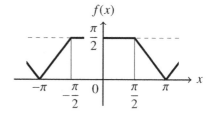

Solution

(a) when $x = 0$

At $x = 0$, $f(x)$ is continuous and equal to $\dfrac{\pi}{2}$.

If we put $x = 0$, $f(x) = \dfrac{\pi}{2}$ into the Fourier series, we get

$$\frac{\pi}{2} = \frac{3\pi}{8} + \frac{2}{\pi}\left(\cos 0 + \frac{1}{3^2}\cos 0 + \cdots\right)$$
$$- \frac{4}{\pi}\left(\frac{1}{2^2}\cos 0 + \frac{1}{6^2}\cos 0 + \cdots\right)$$

$$\frac{\pi}{2} - \frac{3\pi}{8} = \frac{2}{\pi}\left(1 + \frac{1}{3^2}(1) + \cdots\right)$$
$$- \frac{4}{\pi}\left(\frac{1}{2^2}(1) + \frac{1}{6^2}(1) + \cdots\right)$$

$$\frac{\pi}{8} = \frac{2}{\pi}\left(1 + \frac{1}{3^2} + \cdots\right)$$
$$- \frac{4}{\pi}\left(\frac{1}{2^2} + \frac{1}{6^2} + \cdots\right)$$

multiply both sides by $\dfrac{\pi}{2}$

$$\boxed{\begin{array}{l}\dfrac{\pi^2}{16} = \left(1 + \dfrac{1}{3^2} + \cdots\right) \\[2mm] \qquad - 2\left(\dfrac{1}{2^2} + \dfrac{1}{6^2} + \cdots\right)\end{array}}$$

(b) when $x = \pi$

At $x = \pi$, $f(x)$ is continuous and equal to 0.

If we put $x = \pi$, $f(x) = 0$ into the Fourier series, we get

$$0 = \frac{3\pi}{8} + \frac{2}{\pi}\left(\cos\pi + \frac{1}{3^2}\cos 3\pi + \cdots\right)$$
$$- \frac{4}{\pi}\left(\frac{1}{2^2}\cos 2\pi + \frac{1}{6^2}\cos 6\pi + \cdots\right)$$

$$-\frac{3\pi}{8} = \frac{2}{\pi}\left((-1) + \frac{1}{3^2}(-1) + \cdots\right)$$
$$- \frac{4}{\pi}\left(\frac{1}{2^2}(1) + \frac{1}{6^2}(1) + \cdots\right)$$

$$\frac{3\pi}{8} = \frac{2}{\pi}\left(1 + \frac{1}{3^2} + \cdots\right)$$
$$+ \frac{4}{\pi}\left(\frac{1}{2^2}(1) + \frac{1}{6^2}(1) + \cdots\right)$$

multiply both sides by $\dfrac{\pi}{2}$

$$\boxed{\frac{3\pi^2}{16} = \left(1 + \frac{1}{3^2} + \cdots\right) + 2\left(\frac{1}{2^2} + \frac{1}{6^2} + \cdots\right)}$$

(c) when $x = \dfrac{\pi}{2}$

At $x = \dfrac{\pi}{2}$, $f(x)$ is continuous and equal to $\dfrac{\pi}{2}$.

If we put $x = \dfrac{\pi}{2}$, $f(x) = \dfrac{\pi}{2}$ into the Fourier series, we get

$$\frac{\pi}{2} = \frac{3\pi}{8} + \frac{2}{\pi}\left(\cos\frac{\pi}{2} + \frac{1}{3^2}\cos 3\frac{\pi}{2} + \cdots\right)$$
$$- \frac{4}{\pi}\left(\frac{1}{2^2}\cos 2\frac{\pi}{2} + \frac{1}{6^2}\cos 6\frac{\pi}{2} + \cdots\right)$$

$$\frac{\pi}{8} = \frac{2}{\pi}\left(\cos\frac{\pi}{2} + \frac{1}{3^2}\cos\frac{3\pi}{2} + \cdots\right)$$
$$- \frac{4}{\pi}\left(\frac{1}{2^2}\cos\frac{2\pi}{2} + \frac{1}{6^2}\cos\frac{6\pi}{2} + \cdots\right)$$

multiply both sides by $\dfrac{\pi}{2}$

$$\frac{\pi^2}{16} = \cos\frac{\pi}{2} + \frac{1}{3^2}\cos\frac{3\pi}{2} + \cdots$$
$$- 2\left(\frac{1}{2^2}\cos\frac{2\pi}{2} + \frac{1}{6^2}\cos\frac{6\pi}{2} + \cdots\right)$$

Recall that

$$\cos\frac{n\pi}{2} = \begin{cases} 0 & n \text{ is odd} \\ -1 & n = 2, 6, 10, \ldots \end{cases}$$

Therefore

$$\frac{\pi^2}{16} = 0 + 0 + \cdots$$
$$- 2\left(\frac{1}{2^2}(-1) + \frac{1}{6^2}(-1) + \cdots\right)$$

$$\frac{\pi^2}{16} = 2\left(\frac{1}{2^2} + \frac{1}{6^2} + \cdots\right)$$

$$\boxed{\frac{\pi^2}{32} = \frac{1}{2^2} + \frac{1}{6^2} + \frac{1}{10^2} + \frac{1}{14^2} + \cdots}$$

Exercise 2

1.

$f(x)$

$$f(x) = \sin x + \frac{1}{3}\sin 3x + \frac{1}{5}\sin 5x + \cdots$$

Solution

(a) when $x = 0$

At $x = 0$, $f(x)$ is discontinuous.

$$\text{top value} = \frac{\pi}{4}$$
$$\text{bottom value} = -\frac{\pi}{4}$$
$$\text{average} = \frac{(\pi/4) + (-\pi/4)}{2}$$
$$= 0$$

Therefore, at $x = 0$, $f(x) = 0$. If we put $x = 0$, $f(x) = 0$ into the Fourier series, we get

$$\pi = \sin 0 + \frac{1}{3}\sin 0 + \frac{1}{5}\sin 0 + \cdots$$

$$0 = 0 + 0 + 0 + \cdots$$

$$0 = 0$$

(b) when $x = \pi$

At $x = \pi$, $f(x)$ is discontinuous.

$$\text{top value} = \frac{\pi}{4}$$

$$\text{bottom value} = -\frac{\pi}{4}$$

$$\text{average} = \frac{(\pi/4) + (-\pi/4)}{2}$$

$$= 0$$

Therefore, at $x = \pi$, $f(x) = 0$. If we put $x = \pi$, $f(x) = 0$ into the Fourier series, we get

$$0 = \sin \pi + \frac{1}{3}\sin 3\pi + \frac{1}{5}\sin 5\pi + \cdots$$

$$0 = 0 + 0 + 0 + \cdots$$

$$0 = 0$$

(c) when $x = \frac{\pi}{2}$

At $x = \frac{\pi}{2}$, $f(x)$ is continuous and equal to $\frac{\pi}{4}$.

If we put $x = \frac{\pi}{2}$, $f(x) = \frac{\pi}{4}$ into the Fourier series, we get

$$\frac{\pi}{4} = \sin \frac{\pi}{2} + \frac{1}{3}\sin \frac{3\pi}{2} + \frac{1}{5}\sin \frac{5\pi}{2} + \cdots$$

Recall that

$$\sin \frac{n\pi}{2} = \begin{cases} 1 & n = 1, 5, 9, \ldots \\ -1 & n = 3, 7, 11, \ldots \end{cases}$$

Therefore

$$\frac{\pi}{4} = 1 + \frac{1}{3}(-1) + \frac{1}{5}(1) + \cdots$$

$$\boxed{\frac{\pi}{4} = 1 - \frac{1}{3} + \frac{1}{5} - \frac{1}{7} + \cdots}$$

2.

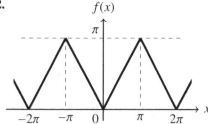

$$f(x) = \frac{\pi}{2} - \frac{4}{\pi}\left(\cos x + \frac{1}{3^2}\cos 3x + \cdots\right)$$

Solution

(a) when $x = 0$

At $x = 0$, $f(x)$ is continuous and equal to 0.

If we put $x = 0$, $f(x) = 0$ into the Fourier series, we get

$$0 = \frac{\pi}{2} - \frac{4}{\pi}\left(\cos 0 + \frac{1}{3^2}\cos 0 + \cdots\right)$$

$$-\frac{\pi}{2} = -\frac{4}{\pi}\left(1 + \frac{1}{3^2}(1) + \cdots\right)$$

$$\boxed{\frac{\pi^2}{8} = 1 + \frac{1}{3^2} + \frac{1}{5^2} + \cdots}$$

(b) when $x = \pi$

At $x = \pi$, $f(x)$ is continuous and equal to π.

If we put $x = \pi$, $f(x) = \pi$ into the Fourier series, we get

$$\pi = \frac{\pi}{2} - \frac{4}{\pi}\left(\cos \pi + \frac{1}{3^2}\cos 3\pi + \cdots\right)$$

$$\frac{\pi}{2} = -\frac{4}{\pi}\left((-1) + \frac{1}{3^2}(-1) + \cdots\right)$$

$$\frac{\pi}{2} = \frac{4}{\pi}\left(1 + \frac{1}{3^2} + \cdots\right)$$

$$\boxed{\frac{\pi^2}{8} = 1 + \frac{1}{3^2} + \frac{1}{5^2} + \cdots}$$

(c) when $x = \frac{\pi}{2}$

At $x = \frac{\pi}{2}$, $f(x)$ is continuous and equal to $\frac{\pi}{2}$.

If we put $x = \dfrac{\pi}{2}$, $f(x) = \dfrac{\pi}{2}$ into the Fourier series, we get

$$\frac{\pi}{2} = \frac{\pi}{2} - \frac{4}{\pi}\left(\cos\frac{\pi}{2} + \frac{1}{3^2}\cos 3\frac{\pi}{2} + \cdots\right)$$

$$0 = -\frac{4}{\pi}\left(\cos\frac{\pi}{2} + \frac{1}{3^2}\cos\frac{3\pi}{2} + \cdots\right)$$

Recall that

$$\cos\frac{n\pi}{2} = 0 \quad n \text{ is odd}$$

Therefore

$$0 = -\frac{4}{\pi}\left(0 + \frac{1}{3^2}(0) + \cdots\right)$$

$$0 = 0$$

3.

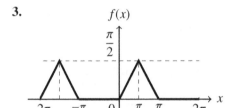

$f(x)$

$$= \frac{\pi}{8} - \frac{4}{\pi}\left(\frac{1}{2^2}\cos 2x + \frac{1}{6^2}\cos 6x + \cdots\right)$$

$$+ \frac{2}{\pi}\left(\sin x - \frac{1}{3^2}\sin 3x + \cdots\right)$$

Solution

(a)   when $x = 0$

At $x = 0$, $f(x)$ is continuous and equal to 0.

If we put $x = 0$, $f(x) = 0$ into the Fourier series, we get

$$0 = \frac{\pi}{8} - \frac{4}{\pi}\left(\frac{1}{2^2}\cos 0 + \frac{1}{6^2}\cos 0 + \cdots\right)$$

$$+ \frac{2}{\pi}\left(\sin 0 - \frac{1}{3^2}\sin 0 + \cdots\right)$$

$$-\frac{\pi}{8} = -\frac{4}{\pi}\left(\frac{1}{2^2}(1) + \frac{1}{6^2}(1) + \cdots\right)$$

$$+ \frac{2}{\pi}\left(0 - 0 + \cdots\right)$$

$$\boxed{\frac{\pi^2}{32} = \frac{1}{2^2} + \frac{1}{6^2} + \frac{1}{10^2} + \cdots}$$

(b)   when $x = \pi$

At $x = \pi$, $f(x)$ is continuous and equal to 0.

If we put $x = \pi$, $f(x) = 0$ into the Fourier series, we get

$$0 = \frac{\pi}{8} - \frac{4}{\pi}\left(\frac{1}{2^2}\cos 2\pi + \frac{1}{6^2}\cos 6\pi + \cdots\right)$$

$$+ \frac{2}{\pi}\left(\sin\pi - \frac{1}{3^2}\sin 3\pi + \cdots\right)$$

$$-\frac{\pi}{8} = -\frac{4}{\pi}\left(\frac{1}{2^2}(1) + \frac{1}{6^2}(1) + \cdots\right)$$

$$+ \frac{2}{\pi}\left(0 - 0 + \cdots\right)$$

$$\boxed{\frac{\pi^2}{32} = \frac{1}{2^2} + \frac{1}{6^2} + \frac{1}{10^2} + \cdots}$$

(c)   when $x = \dfrac{\pi}{2}$

At $x = \dfrac{\pi}{2}$, $f(x)$ is continuous and equal to $\dfrac{\pi}{2}$.

If we put $x = \dfrac{\pi}{2}$, $f(x) = \dfrac{\pi}{2}$ into the Fourier series, we get

$$\frac{\pi}{2} = \frac{\pi}{8} - \frac{4}{\pi}\left(\frac{1}{2^2}\cos 2\frac{\pi}{2} + \frac{1}{6^2}\cos 6\frac{\pi}{2} + \cdots\right)$$

$$+ \frac{2}{\pi}\left(\sin\frac{\pi}{2} - \frac{1}{3^2}\sin 3\frac{\pi}{2} + \cdots\right)$$

$$\frac{3\pi}{8} = -\frac{4}{\pi}\left(\frac{1}{2^2}\cos\frac{2\pi}{2} + \frac{1}{6^2}\cos\frac{6\pi}{2} + \cdots\right)$$

$$+ \frac{2}{\pi}\left(\sin\frac{\pi}{2} - \frac{1}{3^2}\sin\frac{3\pi}{2} + \cdots\right)$$

multiply both sides by $\dfrac{\pi}{2}$

$$\frac{3\pi^2}{16} = -2\left(\frac{1}{2^2}\cos\frac{2\pi}{2} + \frac{1}{6^2}\cos\frac{6\pi}{2} + \cdots\right)$$

$$+ \sin\frac{\pi}{2} - \frac{1}{3^2}\sin\frac{3\pi}{2} + \cdots$$

Recall that

$$\sin\frac{n\pi}{2} = \begin{cases} 1 & n = 1, 5, 9, \ldots \\ -1 & n = 3, 7, 11, \ldots \end{cases}$$

$$\cos \frac{n\pi}{2} = \left\{ -1 \quad n = 2, 6, 10, \ldots \right.$$

Therefore

$$\frac{3\pi^2}{16} = -2\left(\frac{1}{2^2}(-1) + \frac{1}{6^2}(-1) + \cdots\right)$$

$$+ 1 - \frac{1}{3^2}(-1) + \cdots$$

$$\boxed{\begin{aligned} \frac{3\pi^2}{16} &= 2\left(\frac{1}{2^2} + \frac{1}{6^2} + \frac{1}{10^2} + \cdots\right) \\ &+ \left(1 + \frac{1}{3^2} + \frac{1}{5^2} + \cdots\right) \end{aligned}}$$

Made in the USA
Middletown, DE
12 August 2023

36596396R00150